They Were Smokin'

Notice

Mention of specific companies, organizations, authorities or competitive BBQ teams in this book does not imply endorsement by the publisher, nor does mention of specific groups imply that they endorse this work.

Various recipes in this book were taken from the public domain and deemed usable by the publisher. The pork and beef cuts charts included were originally printed by the U.S. Department of Agriculture.

Printed in the United States of America by
Book Masters, Inc.
Ashland, Ohio

Billy Bob Billy
They Were Smokin': Championship BBQ Tips and Techniques from Two Time Memphis in May World Champions Holy Smokers Too / Billy Bob Billy

ISBN-13 **978-0-61523-103-7**
ISBN–10 **0-61523-103-9**

Second Printing 2009

Contents

Dedication

The triumphs and tragedies of a quarter century of competitive BBQ would not have occurred, much less been put to written form, if not for three forces in my life.

First my wife, Susan (also known as Suzie Que) and my family poked fun at me each year this cookbook was not complete. Their motivational style was typical of a Morris and, in the end, effective. My children virtually grew up on the BBQ competition circuit; especially Lauri, the baby of the family. Melissa and Lauri both indoctrinated their husbands Scott (Scooter) and Reid (Action Jackson) into the BBQ scene at many a Memphis in May competition. Lauri picked up the nickname Grazer Jr. after she downed two racks of ribs offered by a friendly competitor in Kennett, Missouri. My son William picked up his old dad's banner to continue my wild times as a young BBQ'er. He later matured as a single dad and brought Billy Bob Billy III (Trey) down to the river as a baby so that three generations were present for several years. Family and friends has always been a key ingredient in this endeavor.

Second, my childhood friend Mike DeVois (Blaze Dawson) made me hang around for several years before he invited me to join the Holy Smokers Too competition team. He knew me well enough to test my commitment before allowing me on the team. He has been a true friend for almost fifty years – some better than others. The part of competition I miss the most is those early morning rides to pick up our hog and the sunrise shifts we shared at many a contest watching the sun come up while smelling the smoke from an army of competitors as we consumed the first Bloody Mary of the day. Thanks, too, to the core group of "old guys" who have outlasted them all...Jimmie (IE), Rock (the original Blaze Dawson), Ernie, John (the original Grazer), also, Rick and Lynn who were there during the *Decade of Dominance*.

And lastly, to Louis Fineberg, who deserves thanks from all the old-time Memphis competitors. Louis was a great supporter of early competitions. He was known around the Mid-South as the "unofficial pork ambassador." He went far beyond his responsibilities as owner and salesman for Fineberg Packing to see that each and every competitor received a quality product. Any tip or piece of knowledge on pork handling and cooking was always yours for the asking from Louis - or sometimes, even if you didn't ask for it. One of Louis' strongest traits was that he always spoke his mind and would tell it straight. He had a dry wit and great sense of humor. He once, personally provided pork shoulders for a team in Missouri who had not received their shoulders for the contest, even though that required driving hours from Kennett, Mo. to Memphis and back again.

"Show me your product," Louis would say. He would then give you his honest opinion and/or pass along a tip on what other teams were doing in competition. Louis left us way too early in life. I wish he was here to give me his opinion on my product.

Louis Fineberg died on March 14, 1995 after a two week struggle following an accident involving the lighting of his charcoal grill using gasoline. He was only 55 years old.

Introduction

My first memories of BBQ were from the city of my birth Union City, Tennessee. (My grand-mother lived in Union City and I was born during a visit). Union City is a small town in northwest Tennessee near the Kentucky line and Reelfoot Lake. In the summer time, the odor of pigs being processed at the Reelfoot Packing Plant sometimes hung in the hot, humid air like the smell of athletic socks in the football locker room during two-a-days, but that's another story.

There was a Pitmaster of note in that small town who had once been called to the White House to dig a pit on the lawn at 1600 Pennsylvania Avenue to serve southern barbecue to the President and other dignitaries. I recall this old, thin black man who tended a large pit dug into the ground covered by corrugated metal sheets he had rescued from a barn that had fallen to the earth. He probably had several hogs or shoulders going at one time, though I only remember him raising a sheet of metal just long enough to remove a large chunk of barbecued pork he then gave to several large black women in sauce-covered aprons for processing. I remember my brother John and I standing at the back door of his barbecue shack like two baby robins, mouths open and head tilted back awaiting a friendly hand bearing pulled pork that was dropped into our waiting mouths. I was five and he was six, but we were both sold on barbecue as one of life's greatest treasures.

Although I have tasted barbecue in many states with many different sauces and presentations, I still prefer the taste of freshly pulled pork, still steaming from the hog's juices and pink with a smoke ring developed over a long slow cooking process. If you ever get the chance to reach in and pull pork from a whole hog or shoulder that's been smoked overnight, take the plunge. Your taste buds will receive a delight impossible to match with pork that has been pulled and/or chopped to be served to you at a later time.

I'm from the South and proud of it. Most of my childhood was spent as a faculty brat at the University of Mississippi. Thirty years of my adult life were spent in Memphis. I mention this only to let you know that when I say barbecue, **I mean pork.** I'm a big fan of Emeril Legasse, but we knew "Pork Ruled" before he served his first mud puppy at Commanders Palace. It's nice to have your basic beliefs confirmed by one of the worlds greatest.

In the early eighties, my best friend, Mike DeVois a.k.a. *Blaze Dawson* began competing in the Memphis in May International BBQ Fest (a.k.a. Memphis in May World Championship BBQ Cooking Contest) *See the history of the Holy Smokers, Too in the section on Competitive BBQ.* I hung around for about five years until someone left the team and they allowed me to join in 1988. It's been a marvelous obsession with pork, hickory smoke, and the camaraderie high-lighted throughout this book.

If you have no interest in tricks, tales, or competitive techniques you can go straight to the recipes. But if you want a good belly laugh, or you have always wondered how and why BBQ teams do what they do, then I recommend you pour yourself a good cold one and settle into your favorite chair or couch and read this book cover to cover before you try to pick a recipe or technique to try.

Confessions of a BBQ Addict

My smokin' name is William Robert Williams a.k.a. Billy Bob Billy and I am a porkaholic and addicted to smokin'. It started the first time I tasted pork while smelling hickory smoke wafting from a backyard smoker. I now have a Pavlovian response whenever I smell aromatic smoke whether it's hickory, mesquite, apple, cherry . . . .and the list goes on. I've even had a "false positive" response to fall leaves burning in the neighborhood. Last holiday season, my family caught me crushing walnut shells and tossing them into the fireplace for a quick fix.

Generally I begin to salivate and search out the source to get relief. I've met some interesting people after climbing their fence to join their picnic in the backyard or neighborhood park. The local police don't even respond when they receive the call regarding an intruder. They just have the caller put me on the phone and remind me of the general restraining order imposed by the county J.P. I'm now able to leave the area on my own without intervention by friends and embarrassed family members.

My last trip from Knoxville to Memphis for the Memphis in May World Championship Bar-B-Q Cooking Contest, normally a 6 hour drive, took me 14 hours. I was lured from I-40 sixteen times and ate six racks of ribs (two baby backs, two spare ribs, one St. Louis cut, and one six pound slab cut from a 350 pound hog headed for the sausage grinder), 6 pulled pork sandwiches, one chopped sandwich, two BBQ plate dinners, and a suckling pig smoked in apple wood that was incredible.

I found the suckling pig being smoked in an improvised pit at the I-40 exit at Buck Snort, Tn. I met two world travelers named Jim and Joe when I pulled off I-40 for a nature call at 2:00 a.m. Jim explained that Joe had climbed a fence into a nearby apple orchard to search for some apples, since they had not stopped their travels long enough to eat for at least two days. Seems they were in a hurry to get home and change out of the striped coveralls they were wearing. I have never liked stripes myself, even if they were U.T. orange.

Joe said he found a good tree and climbed up to pick some apples. He had reached the top and filled his pockets when he encountered a possum hanging by its tail and munching on apples. He panicked and took a swing at the possum, lost his balance and grabbed for the branch holding the possum. It was an old rotten branch so Joe, the branch, and the possum tumbled to the ground. Jim heard Joe hollering and came to his assistance. Jim believes Joe could have been killed, but the possum broke his fall. Jim said they picked up the apples and used the broken branch to smoke the fine BBQ they were willing to share with me. They never did tell me what happened to the possum or where they found the small pig they were smokin'.

It was after I relayed this story to my family and friends that they suggested I seek professional help. It is on the advice of my therapist that I have chosen to assemble the stories and recipes in this book. She says that writing down my thoughts and experiences will allow me to get control of my obsession with BBQ and talking with others with similar interests will help me to accept my problem as one which seems to affect more and more people each year (there were over 100,000 at this year's Memphis in May).

I want to thank all of those fellow BBQ'ers who tolerated my questions and shared their tips, tales, and techniques with me as part of my therapy. If I have misquoted anyone or failed to give credit were due, please forgive me. My therapist suggests I work on one problem at a time.

History of BBQ

Ever since primitive man came upon the cooked carcass of an animal caught in a wildfire man has understood the benefit of meat cooked by fire. Some BBQ offerings still appear to have been cooked by accident, but others have achieved a level of culinary excellence unmatched by other food preparation methods. My hope is that by the end of this book your results will approach the latter, or at least raise to the level of neighborhood bragging rights.

The term Bar-B-Que can be traced back to the French phrase *barbe a que* "whiskers to tail" referring to cooking the whole animal over a pit. Some historians trace bar-b-que origins to the Spaniards in central Mexico where barbacoa refers to lamb or goat wrapped in maguey leaves and roasted on hot coals. In Texas, Mexican vaqueros used this method for cow heads. From the Aussies who "throw a shrimp on the barbee" to the backyards across the USA or the tailgating ritual on any given weekend, everyone has their own view of how to cook over an open flame.

But in this book, we will focus on that long slow-cooked method using charcoal and wood smoke to flavor the meats, and some other surprising offerings. This is not to say that you can't add smoke flavor to grilled fare, some of which we will touch on. But generally, we're talking about a true southern ritual which allows for the consumption of at least two or more cold beverages during the process.

Whether your cooker is a pit dug in the yard or built of cinder blocks and topped with a discarded mattress spring set, a Webber kettle, 55 gallon drum, professional smoker or, in my case, two bathtubs attached by hinges in clamshell fashion, the secret to success is in the process - not the equipment you choose to use.

Americans love to win. Backyard cookers have always enjoyed sharing their success with friends and neighbors. And, as you might expect, in the search for bragging rights, sharing of recipes and techniques had developed into a friendly neighborhood competition. That is until the early 80's when the first BBQ contest was held in Memphis, TN. (Covington, TN claims to have the oldest continuous BBQ contest). That event has exploded from a small group of a dozen competitors to an international BBQ festival lasting nearly a week with teams from around the world. In addition, other contest networks have been formed across the country. In this book we will discuss rules, regulations, tips and techniques for competing to win in any of the various contests in your area.

In Texas, BBQ is beef, often sliced with BBQ sauce added. The Kansas City BBQ Society insists on pork shoulder, pork ribs, chicken, and beef brisket. In Memphis, the Pig is King, although the "Anything Butt" competition is open to your imagination. The West tends to have a fiery tomato based sauce, Memphis and the Mid-South prefer a sweeter tomato based sauce, and as you move east into the Carolinas you'll find vinegar based sauces in the north and mustard based sauces in the south. They all do it "their way" and they all do it right.

Ed Mitchell, the King of Whole Hog cooking in North Carolina, waxed poetic about the history of hog smoking and the roots of North Carolina BBQ. He stated that the hog was never meant to be cut up into pieces, that the 252 distinct flavors of the hog should be enjoyed as one. Ed is famous for crisping up the skin of his hogs after he has pulled the meat. He then chops the cracklin' skin in with his chopped hog meat. The end result is a truly remarkable feast. But holding on to tradition in the face of changing tastes and access to varied new spices and new tools and techniques ignores today's BBQ.

As a twenty-five year veteran of whole hog competition I would have to respectfully disagree with Ed. To chop a whole hog into a mélange of meat and skin, plus significant fat and gristle, ignores the benefit of savoring the many flavors of the unique parts of a whole hog: the wonderfully smoky tastes that differ from shoulder to ham, the buttery flavor from the tenderloin and loin pieces; the tasty strips of bacon and the delicacy of pulled jowl meat, only to name a few. It's as if you prefer a good hamburger made from the whole cow rather than a good sirloin, fillet mignon, or standing rib roast.

So, let's take the trip through the following pages and give you the information you need to **perfect your way of cooking the best BBQ in the world.**

Smokin' Equipment

Cookers

Pit cooking in the ground can yield a sweet, succulent product as a result of the consistent heat-holding characteristics of mother earth. First, over a few beers, talk your neighbor into digging the pit on his property. "You dig the pit and I'll buy the meat." The wisdom of this decision will become evident when we reach the end of our discussion. Pick a spot a distance from any structure and away from overhanging limbs. Otherwise, you may have a dozen additional guests from Fire Station 13. The size of the pit will depend on two factors: (1) how much meat are you going to cook and (2) what do you have available to serve as a rack. If you're cooking a whole hog, then a 3 x 5 foot pit should be large enough. Dig the pit to at least two feet deep. You only need about a foot for cooking, but again, your neighbor will thank you when we get to the end. If you're using a preformed rack – a piece of steel grid, spring set from an old bed, etc. then, be sure and leave a six inch access space along one side to add charcoal, if needed. Trust me, lifting a 125 pound (or heavier) hog on a hot rack to add charcoal is difficult and dangerous, even for four guys who haven't been drinking.

If you don't have an appropriate rack, steel bars or pipes spaced at four inches, imbedded in the ground about a foot from the bottom of the pit and covered with chicken wire should work just fine. So, now you have approximately twelve inches for your fire box and twelve inches above your rack to hold your meat. That extra space above the rack will make it easier to cover the pit with sheet metal or plywood. The sheet metal will get hot, but bends more easily. If using plywood, use the ¼ inch 4x8 sheet and pre-soak it for an hour before using. It will be fairly flexible but may need to be sprayed with water on an hourly basis if your fire gets too hot. Combustion should not be a problem if you keep your fire in the 220-250 degree range with wood smoke retarding the flames. You can use the poor-man's smoker solution, which actually makes clean-up easier. Buy a roll of commercial, heavy-duty aluminum foil and make a cover by crimping sheets together. It's lightweight, durable and can be stuffed in a garbage bag when finished.

Any of the cooking processes and recipes in this book can be used in your pit.

When clean-up time comes, simply remove all metal from the pit, hose down the charcoal, and fill it in with the dirt you reserved when you dug your pit. *You did reserve the dirt on a plastic sheet so you could easily fill the hole and replace the sod, right*! If you got lazy and only dug down one foot for your pit, you can count on every dog and raccoon in a ½ mile radius visiting this spot until you exhume your pit for next year's BBQ. If you don't believe me, then ask my buddy J.J. Butz in Charlotte, N.C. Better yet, ask his wife, Colleen – she'll tell you the truth. He's had late night visitors in his yard for the last several years.

Above-ground pits can produce the same cooking experience. Normally, they're constructed of cinder blocks to the same dimensions as above. Give yourself a foot for a fire box, six inch access, and a foot or more for your meat and cover. But remember, for clean up you better form a multiple layered floor of heavy duty aluminum foil or the critters mentioned above will be nightly visitors to this spot for months to come. If this is a semi-permanent spot for the summer, be prepared for the occasional night time visitor as long as the smell of cooking juices lingers.

Smokers

Smokers come in all shapes and sizes. I've seen a Texas rig so long you good start a hog running in one end and he'd be ready for eating by the time he reached the other end. On the other hand, most backyard cookers are a variation of the omnipresent Weber kettle cooker or the number one homemade smoker crafted from a recycled 55 gallon drum.

Key elements to look for in a smoker are:

(1) Thickness of material – the heavier the metal the better heat retention. In addition, time and the elements are less likely to rust out heavy metal cookers - like so many thin metal cookers that need to be replaced every so often. A well made, heavy duty cooker should last a lifetime or even pass down a generation or two. To prevent exterior rust, opt for a baked-on rather than a sprayed-on finish.

(2) Cooking surface – Obviously, the more cooking surface the more product you can deliver. Newer cookers have multiple shelving systems so you can optimize space in the cooking chamber. Removable shelves are also easier to maintain and make cleaning the cooker much easier. Newer models are available with porcelain racks and cooking surface that make clean up a snap and reduce food sticking.

(3) Fireboxes should allow easy access for building your fire and for the addition of charcoal during extending cooking sessions. There should also be easy access for adding wood chips during the process with minimal loss of heat. Equally as important, there should be a clean-out door or slot to allow removal of ash and cooking byproducts to keep your cooker sanitary as well as prevent rust-out from caustic wet ash left over an extended period of time. Offset fireboxes attached to one end of the cooker allow easy access, but the cooker temp can vary 10 – 15 degrees across the cooker requiring you to reposition meat during the cooking process. My favorite Lange smoker features a heat dispersion plate that forces the heat and smoke to travel the full length of the smoker before entering the chamber and then doubling back to exit the smoke stack near the fire box.

(4) Smokestacks should have a closable flap or baffle to assist in smoke and heat control. With offset fire boxes, the smokestack should be at the opposite end of the cooker to promote heat and smoke distribution across the entire cooker. Larger cookers should have multiple smokestacks, if needed, in order to assure even smoke movement within the smoker and to control heat across the pit.

(5) Air Vents should be adjustable and located near your fire source. Control of air flow, necessary for combustion, is critical to managing your cooker's heat.

(6) A reliable thermometer is essential to monitor the cooking process. If your cooker does not have one, a candy thermometer can be inserted in an air vent on top of the cooker. Better yet, purchase one with an easy to read dial at your local Home Depot, and install it in your cooker. If you don't know what your heat is, how can you know what your meat is doing in the cooker?

(7) Work surfaces attached to the front or side of your cooker are extremely handy and help you avoid dragging your card table out to the cooker for work space.

(8) Hooks or hanging bars for your cooking tools also make it handy to have what you need at your fingertips during the cooking process. A removable magnetic knife and tool bar works well if you have a flat surface available.

(9) Supports and wheels should be heavy duty. Welded legs and supports are preferable to those that are bolted on and tend to loosen over time.

NOTES

Equipment Maintenance

It should go without saying that maintaining and cleaning your equipment is as important to a good product result as your cooking process. But, I'll make a few comments anyway.

If you have just purchased, built, or otherwise finagled a smoker for your use, it needs to be properly cleaned and seasoned. The same holds true as you start a new season or have had your smoker stored for any length of time.

For your typical backyard smoker, cleaning may be as simple as a good grill cleaning with a wire brush and application of a grease cutter on the inner chamber and outside surfaces. Check to make sure all vents are opening and closing easily and look for insects who have built a home in your smoke stacks. Rust spots should be brushed with wire brushes or steel wool and painted with high heat resistant paint. For smokers with an offset fire box, you may find that it is impossible for paint to last due to the intense heat. In that case, I find that coating the outside of the fire box with a vegetable oil spray will hold off the rust and develop an acceptable patina on your box's surface. Vegetable oil sprayed on smoker hinges and vent flaps will also make them work more easily during cooking.

After a good brushing of the grill surface, you should re-season your grill just as you did when you first acquired it. *You did season your smoker when you brought it home from the store, right?* The seasoning process is just like you would use for your cast iron cookware. If your smoker is s store-bought variety with a painted enamel cooking chamber, those surfaces don't need to be seasoned until your use has begun to remove the paint from the metal surfaces, but an occasional light coating of vegetable oil may help the paint remain flexible and less likely to crack and peel.

Seasoning your smoker is a straightforward process. Start a fire and bring the smoker to a temperature above 350 degrees. At this point, take your bottle sprayer and spray water on the grill surfaces while rubbing with your wire brush for one final cleaning. Be sure and wear your heavy rubber gloves for this process since the heat will convert the water to steam when it hits the hot surfaces. Also spray the other inside surfaces for a final level of sterilization. Let the surfaces dry and then apply a coating of vegetable oil on all surfaces with your vegetable oil spray bottle. Maintain the 350 degree temperature for at least one hour. This "seasoned" coating will inhibit rust and reduce sticking during the cooking process. I reapply a thin coating of oil to my cooking surfaces each time I fire up the smoker. Finally, I recommend that you clean your grill surfaces after each cooking session while the grill is hot and then before each cooking session to remove any tidbits you missed from the last use and any foreign material that may have made it onto your smoker since the last cleaning. I know that after a long day of smokin' and drinking cold beverages you may not be as diligent at cleaning your equipment as required. That's why I recommend the second, pre-use, cleaning.

For our competition smoker, Holy Smokers Too always disassembled the smoker and used commercial degreasers and a pressure washer to clean the equipment. This is a lot of work and may be overkill for your backyard needs, but when you are serving the public, you can't be too careful with food safety issues. You would be amazed at the smokers I've seen next to us in competition that are not cleaned between contests. Some containing amazing fungi and bacteria growths any biologist would find intriguing. The teams usually say, "It's O.K., we'll just burn it off when we crank up our smoker." Needless to say, we do not swap BBQ with those teams.

NOTES

BBQ Tools

Gloves are essential to prevent burns, manage the cooking process, as well as ensure a sanitary product for your guests.

Heavy leather gloves come in handy for handling and positioning charcoal, lifting hot cooking grates, or opening and closing vents and smoke stacks. There are some good, and expensive, heat resistant gloves on the market. I prefer heavy duty leather work gloves. They offer good protection and when they become so soaked with grease as to be flammable they can be inexpensively replaced.

Rubberized gloves are essential for handling meat during the cooking process as well as pulling pork prior to serving. Some rubberized gloves have insulation, but the key is to keep steaming juices away from your skin. Many an experienced cooker has performed a Richard Pryor sprint down the road trying to remove a glove soaked in boiling pig fat. Gloves that reach to your elbow will protect your forearms from a chance encounter with hot pork or the sides of your grill. If you have a cooker that allows you to add charcoal and chips from underneath the cooking surface, this will also protect you from drippings from above.

Latex gloves are a must in the final serving process. Your guests appreciate your cooking efforts, but don't want to share the sweat and grime from your cooking experience. Alcohol based hand sanitizers in a spray bottle are also recommended for your list.

Long handle spatulas, tongs and brushes will also provide you the ability to tend your meat without exposing your hands to flames and heat. You may have noticed I omitted fork. Forks are for eating – not cooking. You do not want to pierce your product unnecessarily during the cooking process thereby releasing valuable juices. Buy the best equipment you can afford. It will last longer and provide the best safety for you. I use a heavy duty set of tongs with scalloped tips that allows me to easily lift a rack of ribs or Boston butt with one hand. Two of the tongs allow me to turn a shoulder.

Wire brushes for cleaning your grill can be quite impressive and expensive. Some have long handles, heads with both wire brushes and steel wool pads. Some have scrapers attached to the ends. Unfortunately, most of them don't last more than one season. I find inexpensive paint scrapers with wire brush and pointed steel scraper, to work the best. The bristles last longer and they are inexpensive to replace.

Spray bottles are a life saver. I always have at least three on my equipment list. (1) A water bottle handy to douse any flare-ups that may occur while I'm tending the meat. (2) An oil bottle with cooking oil to spray on the cooking surface before cooking, to help with cleaning, to reduce sticking and during clean-up to spray a coating on the cooking surface to reduce rusting. (3) A bottle with basting or other moisturizing solutions to spray on the product if it's thin enough for a spray nozzle.

Shovel and rake for adding charcoal and tending the fire. A garden hoe works well for moving ash away from the fire or a 4 inch weed rake works well in pulling live coals to the surface away from built up ash. You can buy inexpensive garden models and simply cut down the handles to around 24 – 36 inches. For your small cookers, hand sized garden tools work just as well.

Rib racks allow you to maximize the number of ribs you can cook at one time. These racks allow you to cook ribs standing on end. This gives you the opportunity to cook three times the ribs you could by cooking them flat. I have a great set I bought from Coleman, the camper people, that are porcelain coated and easy to clean. They also work well for chicken breasts and trimmed beef brisket.

Meat thermometers are necessary to accurately judge the doneness of your product. Instant read digital thermometers are economically available in many stores. The new remote thermometers allow you to insert a probe into your meat and monitor its internal temperature without raising the door of your cooker. They will even beep you up to 100 feet away when your meat reaches your target temperature - a feature you may not want to use if the beer's at your side and there's too much work going on in the house.

Rib stripping tools are part of the mystique of the competitive world. Once identified, they are fairly easy to come by, but are absolutely required to properly prepare Championship Baby Back Ribs. It's a secret I'm not ready to reveal at this time. Should you actually work your way to the rib cooking section you will find the answer.

Clamp-on lights or a headlamp are good add-ons for your tool list if you find yourself cooking at night without adequate light. They will help you check your cooker without an assistant to hold a light.

Household fire extinguishers might be the last, but most important item on your list. Accidents happen when you least expect them to. Take the risk seriously and place an extinguisher near your cooker. Nothing ruins the ambiance of your cookout like the smell of burning hair - yours!

Basting brushes or mops are also a matter of personal preference usually controlled by the meat you will be cooking. For light coats of baste, olive oil or butter the typical long handled paint-brush-style basting brush works well. For sops and bastes where you wish to carry a large quantity of liquid from bowl to meat I prefer the mops that actually look like small versions of your typical floor mop. One inch wide strips of towel or t-shirt stapled to a wooden handle work quite well also. I like to don my rubber gloves and use the towel strips without the handle. This allows me to squeeze more baste in spots where it's needed.

Fire & Smoke

When we talk about slow smoking bar-b-cue, it's generally understood on this side of the Mason-Dixon that we are talking about cooking over a charcoal fire. You can make your own charcoal or choose from a wide range of products on the market today.

If you find yourself with an abundant source of hardwood and the time and patience required, you can make your own charcoal. Simply arrange your wood so as to create a good fire that will burn the original wood down to red hot coals. Spread the coals in your pit or transfer them to your cooker and begin the cooking process. The most common choices for charcoal are hickory, oak and mesquite, but the addition of fruitwoods such as cherry and apple, or pecan can add interesting flavors. Soft woods like the abundant southern pine are to be avoided unless you just like the taste and smell of paint thinner. The resins in soft wood are pungent and tend to coat your cooker and your meat with a sticky coat.

Aged firewood or wood sticks from hardwoods can be used as a heat source in offset fireboxes or at the opposite end of a large smoker that allows enough room between the heat source and the meat so that flare ups and charring from radiant hit does not occur. Remember, we want to cook with heated air and not from the radiant heat from flames or glowing coals.

Commercially available charcoal generally comes in either lump or briquette form. Purists prefer lump charcoal because it does not contain the fillers and binders that briquettes may contain. Lump charcoal also ignites easily without the use of accelerants. Briquettes are made from lump charcoal that has been pulverized and then compressed into the briquettes you buy at the store. Fillers made of sawdust from other woods as well as a binding element are added to form the briquettes. I prefer the briquettes because their uniformity in size and heat allows me to direct and control the heat. High quality briquettes tend to leave a minimum amount of ash when burned. I have yet to talk to a competitor who expressed a concern about taste being compromised by the fillers found in briquettes. Those who do will also tell you they can taste the bourbon used to soak the hickory chips. (See the section on smoke.)

Some teams use a gas starter to ignite their charcoal. Holy Smokers Too now uses a brush burner attached to a propane tank. It's quick, efficient and does not require the use of petrochemicals. If you choose to use a liquid charcoal starter, buy a national brand such as Gulflite or Kingsford. Some cheaper fluids don't burn off as readily and leave a lingering odor/taste.

To start your fire with liquid starter you should first build a mound of charcoal briquettes in your firebox or offsite box. Saturate the briquettes with starter. Don't overdo it! Typically one cup of fluid is more than enough for 20-25 briquettes. Remember, you're just starting a fire – not building one to cook on. Wait a minute or two so that that the fluid has time to soak into the briquettes, otherwise you'll just burn the fluid off the surface.   With all the cooker's vents open, light the fire and then wait until all of the briquettes are covered with a grey ash.

Charcoal chimneys are very effective in starting your fire. You simply load the top with your charcoal, crumple a newspaper under the chimney, ignite it and in as few as fifteen minutes you will have a kettle of red hot coals ready to add to your cooker. You can also use a liquid or gel charcoal starter in a chimney to start your fire.

There are also many gel based fire starters or starter sticks that can be used to ignite a fire by placing the gel cubes or sticks at the base of your charcoal and following your usual ignition sequence. Find a method that works well for you and stick to it. Save your creativity for the cooking process.

Never, and I mean never, use gasoline to start this or any fire. That's exactly how we lost Louis Fineberg, for whom I have dedicated this book, and serves as a real-world reminder of the dangers of accelerants. Also, do not pour, or squirt, additional fluid on a slow fire or smoldering briquettes. In addition to a flare up that can quickly remove body hair or light your clothing, squirt cans may also suck a flame back into the can with explosive consequences. If you must increase a fires progress after the initial lighting, use a fan or increase the draft by adjusting the vents. In addition, you can start additional charcoal in your chimney or a separate spot in the cooker and combine these coals with the original when they are hot. *Invest in a household fire extinguisher and keep it handy.*

And when you inadvertently come in contact with a hot surface, apply ice immediately to stop the cooking process. It works for the vegetables you want to keep crisp by stopping the cooking and ice has prevented many a burned finger from blistering into a debilitating sore that ruins your cookout. It may take 30 minutes to an hour, or more, but keep applying the ice until the painful tingling subsides.

NOTES

Smoke Flavor

Since charcoal used in cooking is made from hardwoods, some smoke flavor will be imparted to your food without the addition of smoking chips or chunks. However, the majority of BBQ'ers will use additional wood products to get the smoke flavor they are looking for.

Hardwood to generate smoke can come from logs, chunks or chips. Each has its own unique characteristics. Wood used to generate smoke is not meant to ignite, but to smolder and generate smoke rather than heat for cooking. This is accomplished by soaking the wood in water for an extended period of time. For logs or chunks, overnight soaking is recommended. The smaller wood chips can be ready for use after one hour of soaking. The logs and chunks will generate smoke for a longer period of time than the chips. The chips tend to produce a greater intensity of smoke, but dry out and burn more readily than chunks. You will want to experiment and find which style you prefer.

I prefer the chips because of the intensity of the smoke and also believe I can better control the heat in the cooking chamber because of the increase in smoke and the water released from the chips.

Soaking the smoking material prevents ignition, but many believe the liquid used can also impart flavor to the meat. I know that the Jack Daniels cooking team uses a heavy dose of a not-to-secret liquid to soak their chips. A sampling of chip buckets might find bourbon, apple cider, cherry juice or any wide range of elixirs teams will swear by. Superstitions run wild thru the competition circuit.

Cooking with gas is not allowed on the competitive circuit, but it may be possible to smoke your 'que on your gas grill if you can keep the cooker under 350 degrees and use a wood box or smoking packet to hold your wood chips. We'll explore that option when we are discussing the cooking process. Some of the newer, or modified, smokers actually have a gas starter used to start the charcoal and in some cases this comes in handy to finish off the smoked cooking of large cuts of meats that cook for over 5 to 6 hours.

Smoked Cooking means many things to many people. I'm about to give you my nickel's worth of advice acquired over the past 20 years.

Smoke has been used for centuries to enhance the cooking process. Early civilizations realized they could preserve meats and fish by applying smoke at temperatures that did not cook the meat but removed moisture from the meat while creating a coating that was not conducive to living organisms that contributed to meat spoilage. Cold smoking at temperatures between 80° to 120° for four to six hours, or longer depending on the thickness of the meat you are smoking, can impart a smoky flavor while acting as a preservative. But, that's a topic for a different book.

The type of smoked cooking we are looking for is meant to enhance the flavor of the meat. The most common wood used in the South is hickory. Mesquite is favored in Texas and parts west of the Mississippi River. Oak probably comes in a close third and then you see a lot of the fruitwoods such as apple, peach, cherry, pecan. Any hardwood will produce usable smoke, but each will impart its own unique flavor to your cooking. Some cooks swear by a mixture of woods to get just the right taste. I've used pecan shells and grape vine cuttings with surprisingly good results. The stronger smoke of hickory and mesquite are fine for beef or pork, but you may find them overpowering for delicate fish or even a cut of lamb.

Wood Smoking Chart

Wood	Characteristics	Recommended
Alders wood	light & aromatic	good with salmon
Apple, Cherry	sweet & fruity	chicken, turkey, pork
Hickory	strong, hearty	chicken, pork, beef
Maple	sweet, mild	chicken, vegetables
Mesquite	strong, tending toward bitter	beef, venison, pork
Oak	strong, hearty	chicken, pork, beef
Pecan	mellow, rich - not heavy	chicken, fish
Grape vines	mild	chicken, fish

Wood for smoking can come in the shape of logs, chunks, chips, vines, sawdust or pellets. Each has different smoking characteristics that may determine your selection. The more surface area the more intense the smoke that's generated. Wood used to generate smoke should be soaked in water, or your secret elixir before use. You want the wood to smolder and produce smoke, but you do not want the wood to produce a flame and add to the cooking fire. Logs or branches from a local tree you have harvested tend to burn slower and produce a lesser degree of smoke than chunks or chips. The major benefit is that you do not have to add wood to the fire as often and the smoke generated holds at a consistent level. Wood chunks will absorb more water, and because of the increased surface exposed to heat will generally generate more smoke than logs. Because of the exposed surface area, chunks may dry out during the cooking process and catch fire. My personal favorite is wood chips. They deliver intense smoke and are easy to handle and add to the fire when needed. Wood chips do tend to dry out more quickly than other forms and will often ignite when the cooker is opened to tend to the meat. But, a handful of wet chips can also be used to stop a flare-up in the cooker.

Finally, because the smoke generated by wet chips is intense and quickly generated when the chips come in contact with hot coals, I use the chips as a part of my fire management and temperature control systems. *See Fire Management.*

The Smoke Ring

The area that generates the most confusion as well as a competitor's greatest pride is the smoke ring visible in smoked cuts of meat. Every pit master points with pride when his pork or brisket shows a pink ring deep into the surface of his finished product. We are often asked how we are able to control the depth of the smoke ring when we cook. The truth is that we don't. There are several factors involved and most are out of our control.

So, what is the science behind cooking with smoke and the smoke ring? There are three component parts to smoke:

(1) Solids – fly ash and tar
(2) Non-condensable – air and combustion gases
(3) Condensable – acids, carbonyls, phenolics & polycyclic hydrocarbons

One and two do little to impart flavor unless you stir up the ash when tending your fire. Then, you can add an unpleasant powdery addition to your meat's surface. I call that "fuzzy barq" or bark by some.

It's the phenolics that interact with the meat's fibers and proteins to produce the smoky aroma and flavor. The smoke flavor is produced from a wide range of compounds found in the variety of woods used in cooking. The density of the smoke in the cooking chamber and the time of exposure will also affect the amount of smoke flavor that adheres to the meat's surface. The carbonyls are the source of the amber color that we call the smoke ring.

There is a limitation to how deep this interaction penetrates into the meat. Normally, the reaction stops between ½ inch to 1 inch into the muscle fibers. The process also begins to diminish as the meat's internal temperature passes 130°.

Louis Fineberg often said you could over-do-it with smoke and end up with a bitter product. He would tell you to put a cup of water in the smoker along with your meat and taste it after several hours to see if it was bitter. I never tested Louis's theory, but I assume the bitterness may have come from ash or bad wood. The theory also does not take into account the counterbalance of barq buildup and the sweetness of basting sauces.

I prefer heavy smoking thru the use of wood chips, but I have judged some fine entries that were cooked using just hickory charcoal, with minimal observable smoke, that produced a great product. That leads some credence to the invisible gases at work.

Blaze Dawson says, "Forget that science stuff, a smoke ring is a smoke ring and that's all there is to it!"

Meat Under Heat – *The difference between BBQ nirvana and disaster.*

> **Read this portion at least three times – until you can recite its contents to your spouse. If you don't have a spouse, don't tell your buddies. They can buy their own copy – let this be your secret.**

*** THIS IS THE SECRET TO GREAT BBQ ***

In BBQ, all of the flavor and nutrients in pork, beef, veal, and other meats comes from the muscle fibers – with the exception for those of you who enjoy the organs themselves (including skin). I know many people make the statement that the flavor is in the fat, but I challenge those people to take a spoonful of lard or Crisco shortening and describe the taste as anything but yuck! The fat does contribute to the juiciness and tenderness of meat, but we will look at that further. Nothing against those of you who still like to burn the fat on your steaks and think it's delicious, but my cardiologist recommends against such a practice.

Muscles consist of bundles of individual fibers held together by connecting tissue. When you look at a piece of meat it's easy to see these fibers as the "grain" of the meat. These bundles are covered with a thin membrane composed of collagen. These membrane sacks also hold fats, natural sugars, salt, vitamins, acid and enzymes. Other sacks hold larger fat deposits. As much as 80% of the contents of these sacks may be water.

As you begin the cooking process, these sack's membranes start to break down. As the sacks release their contents, the water, vitamins, salts, sugars and proteins mix with the melting fat to form the juices in the meat. This transformation occurs in the 120° range. From 120° to 130° of internal temperature the muscle fibers begin to contract. This forces some of the juices to the surface – a condition some cooks refer to as *sweating the meat.*

At 140° of internal temperature the collagen in the tissue breaks down into gelatin. When combined with the juices, this contributes to the tenderness of a good cut of meat. But, at 145° the proteins begin to harden and the meat starts to get a little chewier. Be warned, 149° is a key temperature, the zone at which muscle proteins begin to harden and become tough. Therefore, lean cuts (loins) should not be cooked beyond this range.

At 160° of internal temperature you have reached the point that the USDA authorities believe pork is safe to eat. A pork shoulder will have the consistency we are all familiar with as a baked ham. BBQ'ers, searching for that wonderful pulled pork or juicy brisket, should know they are almost there at this point. Although North Carolina boys have been known to pull out their cleavers and chop 160° shoulders into Carolina BBQ Hash!

At 170° of internal temperature beef is considered well-done. A significant amount of water loss has taken place and the muscle fibers have contracted to their fullest. This is why you may experience a dry, tough cut of meat.

As you raise the internal temperature from 160° to 180°, two critical events take place: (1) the meat loses an additional 30% of its liquid; and (2) the muscle fibers begin to breakdown. The speed of breakdown and the quantity of moisture loss will vary depending on the temperature you are using to smoke your meat.

Somewhere in the 180° to 195° range is where you'll find most BBQ'ers are ready to pull their pork or slice their briskets. Generally, those of us who wrap their meats in the final stages will allow the meat to rest 30 minutes to an hour to allow the juices to redistribute throughout the piece of meat before serving. The cooling of the juices also makes them less runny and less likely to bleed from the meat when cut.

The majority of tough pork shoulders, ribs and brisket can be blamed on not taking your product to this final stage where it once again becomes tender. Even an older, tougher cut of meat can be tenderized between 180° to 185° over an extended period of time.

Beyond 200° your pork will start to turn to mush and beef starts to crumble. This is typical "pot roast" fashion and is past the point of good pulled pork BBQ. Maybe great for a Sloppy Joe.

Meat Changes While Cooking

Internal Temperature	Tenderness/Texture	Juiciness	Color	Micro-organisms
120 Degrees	Muscle fibers begin to contract. Juices are forced to surface. Meat begins to "sweat".	Slight water loss	Beef - bright red Pork - grayish pink Veal - pink Lamb pink/red	Most microbes still actively reproducing
130 Degrees	Muscle fibers are drawn together as connective tissue breaks down and more juices are lost.	Proteins begin breakdown Tender cuts of meat are juicy at this point		Trichinosis culprits killed at 137 degrees
140 Degrees	Rare Collagen breaks down into gelatin - meat is tender to the bite.	Beef is tender and juicy Pork not quite done here	Beef - bright red center Pork - pink center	Yeast's and molds destroyed
145 Degrees	Medium Rare Proteins begin to harden	Still juicy but losing water	Beef - pink center Pork - white center	Pathogenic bacteria are destroyed at 149 degrees
160 Degrees	Medium Federal guidelines for most prepared meats for public.	Less juicy	Beef - pink/gray center Pork - white center Veal - grayish brown center	pork/veal/chicken federal guidelines Reheated foods should go to 165.
170 Degrees	Well Done Meat has completed shrinking Relatively tough chew	Dry	Beef - gray/brown center	
180 - 185 Degrees	Muscle tissue breaks down Pulled Pork BBQ Brisket sliced when cooled to 150	30% of water loss from 160 to 180 degrees. Slow smoked BBQ should be wrapped in foil during this stage	Beef & Pork grayish brown center with pinkish smoke ring.	
200 + Degrees	Overdone - Muscle is mushy			

"If you can't pull it, don't eat it, "
Blaze Dawson.

Wrapping for Rehydration

Many BBQ'ers will wrap their meat in aluminum foil during the final stages of the cooking process. Some consider this technique a lazy alternative to proper care of your product during long hours of cooking. I'm in favor of wrapping to rehydrate and/or preserve juices in my products. The process of taking pork shoulder from 160° to 195° will render 30% of the remaining liquids from the meat. To demonstrate the liquids lost in the final stages, I will place Boston butts or shoulders in aluminum pans for the last stage. Students attending my cooking classes at the University of Tennessee are always amazed at the level of liquids in the bottom of the pans! Try it yourself and you'll need no other convincing of the need to protect your meat in this last stage of the smoking process.

> *As much as 30% of a shoulder's liquids can be lost in the final cooking stages!*

After all, there is a point of diminishing returns where the meat's smoke absorption is concerned. I will wrap ribs after two hours and larger cuts like butts, shoulders and briskets after six hours. There may be additional flavor imparted beyond these times, but the moisture loss is too great a sacrifice for most backyard cooks. Many professionals do not wrap their meat and are able to maintain moist products thru constant basting and low cooking temperatures, or with the addition of moisture into the cooking chamber. Most beginners don't have the patience or experience to avoid drying out their meat.

For long cooking times, you can place your wrapped meat in an oven at 225° for the final stage. This takes some of the fun out of the process, but may offer some much needed relief for the cook. And besides, **"The pig won't know he's not still in your smoker!"**

If you like a crisp barq for your pork or beef, you should remove the meat from the foil for the final cooking stage. Place your meat in the smoker and increase the heat to the 275° range for approximately 30 minutes to dry out the surface and crisp-up the barq. This short period of increased heat will not adversely affect the tenderness of your product.

Remember, we are not talking about grilling fine cuts of meat over high heat here. We are talking about smoking BBQ at 200° to 275°. Low and slow is our mantra.

Fire Management

If you have reached this point faithfully, by that I mean you have read all of the preceding pages, you may be asking yourself, "How can I cook like the big boys when all I have is a Weber kettle?" There are a lot of sophisticated smokers on the market. Some are very large and have automated fire and fuel controls that allow you to "set it and forget it." Without looking at those smokers and their capabilities, we will discuss how to get the same results with less. Besides, as a well known national weatherman discovered, the most expensive, automated smoker in the country is not much good if your power goes down in a Memphis in May tornado.

> **"Besides," says Action Jackson,**
>
> **"It's not the size of your smoker...it's how you rub your meat!"**

Remember, we're talking about cooking "low and slow" with heated air and smoke. Cold smoking of foods takes place below 200°, usually around 160° or less. Grilling, on the other hand, takes place above 400° and uses radiant heat to cook the food. BBQ'ers want to cook in the 225° to 350° range using only circulating air and smoke without the cooking caused by radiant heat. My target rate is always 225° on my smoker.

The intensity of your fire is totally dependent on the fire's access to fuel. The two fuel sources involved are wood products and oxygen. The extent to which you increase or decrease the amount of fuel your fire can consume will control the temperature in your smoker. Other factors such as outside air temperature, sunshine on your smoker, thickness of your smoker's material, etc. all have some impact on your fire. However, fuel is paramount.

If you have an offset firebox, you need only to build a fire of sufficient size to heat the cooking chamber to the desired temperature and keep the fire going, with the addition of hardwood chips or chunks to generate smoke for flavoring.

Indian's Fire or White Man's fire?

The story goes that the Indians knew how to build small fires and gather around closely for heat. This conserved fuel sources and did not alert enemies to their whereabouts. On the other hand, the white man builds a large fire and backs up from the heat, wasting fuel and alerting anyone for miles around of his presence.

In smoking, it is sometimes more prudent to build a white man's fire. If you load up your fire box with charcoal or wood with enough fuel for a bonfire but you limit the amount of oxygen by closing the vents, it is possible to sustain the needed heat over a longer period of time without having to add charcoal or wood.

This may be more difficult if you have a small cooker and have to build the fire within the cooking chamber. Most cookbooks will tell you to build your fire on one side of the smoker and place your meat on the other side for "indirect cooking". Or they may recommend placing your meat in the middle of the smoker over a drip pan with fire on either side. Either method may work, if you are cooking a limited quantity of meat. Also, remember that each time you open your smoker to baste your meat, the air rushing in may cause flare-ups that require the use of a water spray bottle to avoid a grease fire.

What do you do if you're cooking for the whole family and need all of the smoker's grill space to spread your meat out for cooking? You don't want the meat too close to the fire or you risk charring the meat's surface from the radiant heat and flames from your fire. Time to use the white man's fire!

Build a pile of charcoal in the center of your grill large enough to cover the entire grill surface when spread out. After the charcoal has reached the stage when all briquettes are covered in grey ash, spread them evenly across the charcoal grate under your grill surface. Add several hands-full of wet smoking chips to your fire and position your grill with meat across the entire surface. Place the cover on the smoker and close down the vents to smother the fire down to your appropriate temperature. With the chamber full of smoke and limited air, the charcoal will smolder and create the required heat without flare ups. If the charcoal is too close to the cooking rack and you begin to see burning on the edges of your meat, you can place your meat on a cookie sheet for protection from the radiant heat. A double thickness of heavy-duty aluminum foil placed under the meat may also work well. A few strategic holes punched in the foil will prevent juices from pooling on the foil and creating a flammable runoff river of fat.

A more effective solution is to use a pizza stone if you are cooking on a kettle style grill. Place the stone in the center of the cooking area so that the smoke will circle around the edges. If you are using a rectangular grill, you can build a surface with ceramic tiles, leaving 2 – 3 inches around the edge for smoke circulation. If you use either of these methods you will need to place your meat on a metal rack (borrow your wife's rack from the roasting pan) so that there is smoke circulation around all surface areas of your product. It also might be wise to cover your stone or tiles with aluminum foil for a quick clean up. You'll also hear from your kids if you screw up their pizza stone.

When it comes time to baste or add charcoal, be aware that your oxygen starved fire will be prone to flare-ups. It's usually best to plan to remove the meat to a nearby surface for basting and closing down the smoker until you are ready to reposition your meat. This will lessen the chance for flare-ups and the associated risks. In this way, you can cook more meat on a small cooker without "grilling" your meat. If this is your process, you should be aware that your cooking times will be extended due to the heat loss during basting. This should also serve as a reminder to heat your basting liquids before applying so that you are not applying cold baste to warm meat.

As you go through the cooking process, it is important to remember that temperature control in your cooking chamber may be counterintuitive. For example, if your smoker is too hot, do not open the lid or the smoke stacks to let the heat out. This in itself may provide more oxygen to the fire and it will just get hotter. Even with an offset fire box, you may just draw hotter air into the chamber. Conversely, if the cooker is too cool, closing down the smoke stacks and the vents to "hold in the heat" does not work either.

Closing down the air vents and the smoke stacks is the quickest way to lower the temperature in your smoker. Choking off the flames or high intensity charcoal will quickly bring the temperature down while keeping the fire under control. The addition of a few hands-full of wet wood chips will quicken the drop in temperature because of the water on the charcoal and the increased smoke (less combustible air) in the chamber.

Remember, oxygen means heat. To increase the cooking temperature, open the vents to let in oxygen to feed the fire. If the chamber still will not rise to your desired temperature, it may be a problem of air flow. Open your smoke stacks to increase air flow and draw more air into the chamber from the fire. As you add more meat mass to your cooker, you may also need to increase air flow to have the same cooking success as with smaller cuts of meat.

My son-in-law was competing in Dillon, Colorado and could not get the smoker to the desired temperature. Because of the altitude and the fact that he was trying to compete as well as cook several Boston butts for 25 of his friends, he was not getting enough hot air into his cooking chamber. We propped open the door to his smoker by ½ inch with a stick and the temperature finally came up to his desired temperature. All a matter of airflow! *That stick continues to be a valuable tool with his new smoker.*

The BBQ still didn't turn out so good – too much meat mass in that small smoker, **but he did win the salsa competition!**

At competitions the real rewards come from the time spent with friends – old and new. A trophy or ribbon is just icing on the cake!

Charcoal purists insist you only add hot coals to your fire during the cooking process to burn off the gases from the binding agents in charcoal briquettes before placing them in your smoker. I've never been able to tell the difference in the final product, unless you're talking about grilling mild meats such as fish over charcoal that has not reached the grey-ash coating stage and are still giving off gases from the accelerants. Experiment with your smoker to determine the best method of adding charcoal that works for you and yields the best, consistent result.

The same goes for smoke. Try chips, chunks, twigs, or logs. Whichever bests suits your equipment, style of smoking, and the results you prefer. I like smoke. I tend to add smoke chips to my fire whenever I don't see smoke coming from my smoker's stack. (Usually on the hour). Your timing will vary depending on the amount of smoke flavor you are looking for as well as the product you are using (chips, chunks, split wood, etc.) Ambient temperature, wind and rain can also impact your cooking process.

From 55 gallon drums, to bathtubs, to $10,000 professional smokers used by amateurs or seasoned professionals you will find a wide variety of preferences. The keys are to practice, take notes, make changes and then, develop your own unique style.

The Meat Selection Process

No Bones About It!

There's no bones about it..."You can't have a winning product unless you start with a good cut of meat," Louis Fineburg, 1988. Thanks to organizations like the National Live Stock and Meat Board and the National Pork Board you can feel secure that you are purchasing safe products. But safe products are not enough to win contests. While cuts of beef are graded as to tenderness and fat content, pork is not.

Many years ago the National Pork Board and pork producers responded to concerns about healthy eating. The result has been the production of a leaner pig – some say almost too lean. That's great for people cooking the leaner cuts of the pig, but not so good for BBQ'ers. Hogs are now raised in very sanitary controlled environments and sent to the packing house at a fairly early age. This produces an average pre-slaughter weight of 200 pounds and yields a 120-140 pound carcass. A 70% yield for a slaughtered hog is a reasonable industry average. That is not to say that slaughtered hogs in the 300 pound range are not available. These hogs tend to be older and have more fat and tougher muscle tissue. Many large hogs are bound for sausage production.

Holy Smokers Too generally chooses a 120 pound dressed hog that has been freshly killed. Before Louis Fineberg died, and the USDA cracked down on rules on access, we used to select our hog from Fineberg's slaughter house inventory. Louis always made sure we got a female with one blue eye and one brown eye and we named her Oink Johnson. In most cases, purchasers have to rely on the staff at the packing house. On rare cases, a competitor shows up with a farm-raised hog of his own, but not too many contests allow non-USDA inspected meat products. What you do in your own back yard is up to you.

Ask your processor to split the back bone for you. They can do this in the plant with a power saw that is quicker and more sanitary than the sledge hammer and axe we used to use. You might consider having the ribs separated from the backbone by slicing down both sides of the backbone as part of this process. This will allow the hog to lay flat as well as expose some of the loin to smoke and make the ribs easier to remove when the cooking process is completed.

Preferred pork shoulders should be in the 16 – 20 pound range. Again, larger shoulders come from older hogs. The shoulders are the front legs of the hog and should include the shoulder muscles and the shank. When the shoulders are cut in two, the upper section is referred to as a Boston butt or roast, and the lower section is then a picnic shoulder. The rear legs are referred to as the ham and may be marketed as a rump portion and shank as well. The rear leg, or ham, is a less complex muscle structure and tends to be more tender and leaner than the shoulder cuts.

Boston butts in the 8 – 10 pound range have become the preferred cuts for pulled pork because they have a good yield and are easier to handle than a full shoulder. They also require less cooking time. Shoulders require 12 – 16 hours, while butts only require 8 - 10 hours. When purchasing pork shoulders the product should be bright red in color without too much visible fat marbling. Cuts with visible blood clots or dark spots should be avoided.

Spare, St. Louis, or Baby Back

Ribs are without doubt the nirvana of BBQ eating. But why are there so many different names for them? Historically, there were two main descriptions for ribs. First, there were the loin ribs that were attached to the loin muscle running down the upper back of the hog. These were preferred by the owner of the farm, plantation, kingdom, etc. The lower portions of the ribs around the pig's belly where the bacon and lard were harvested were the leftover or "spare" ribs.

When the lord of the manner had eaten all the best parts, living "high on the hog", then and only then was it decided that the left-over parts could be "spared" for the working class. *A trained meat cutter will tell you that the term "spare" refers to the fact that little meat is left on the ribs during processing. Webster also defines spare as meaning "thin or scanty."* I like my story better.

The loin ribs are the same rib bones you would see attached to a bone-in pork chop. As the pork loin gained acceptance as a stand-alone pork cut, the ribs, once detached, gained popularity since they include portions of the loin meat that is left attached. Some smart marketer, decided to start calling these "baby back" ribs to accentuate their tenderness.

Baby back ribs <u>do not</u> come from baby pigs!

Amazingly, as baby back ribs have gained in popularity, they are now more expensive than they were as a pork chop! Not bad for a piece of waste material left over from harvesting the pork loin. As baby backs have become more expensive, we began to see a lot more "pseudo baby back ribs."

The spare rib includes the portion of the rib from mid-rib down to the connecting bones of the chest. Often included is a brisket flap from the chest muscle. If the brisket portion and the breastbone cartilage are removed, along with some fine tuning, you have a clean rack of ribs known as a St. Louis cut (pseudo baby back rib). You can separate the brisket portion by running a knife down the joint between the brisket portion and the main ribs. The leftover brisket portion will contain substantial meat portions, but interspersed with cartilage, fat, and short bones. I refer to that portion as the "chef's rewards" to be eaten while the remainder of the ribs are finished cooking. These ribs will contain more fat than the baby backs, but also yield a lot of meat at a much lower price than the baby back ribs. Some competitors prefer the spare ribs because they don't dry out as easily, but as contests have evolved, spare ribs have a hard time competing with baby back ribs.

Country cut ribs are usually seen as short pieces of bone and pork and come from the last four ribs towards the neck. They are a great buy for your backyard smoker or grill. They are flavorful and will cook in a much shorter time in your smoker or on a grill.

A rack of baby back ribs or spare ribs should contain 12-13 bones. We used to compete with 1.75 and down baby back ribs. That means each rack weighed around 1.75 pounds. Nowadays most baby backs you find are more likely 2.25 – 2.50 pounds. Anything over 2.50 pounds should be avoided for competition since they come from older hogs (tougher and more fat). In addition, the bones should be round, tending toward elliptical. Flat ribs also indicate an older hog. Spare ribs will run 2.75 pounds and up! Remember, heavier ribs mean more fat and an older hog.

When I want to go on-the-cheap to feed a large crowd, I'll buy rib strips from the local sausage maker. These are generally 3 - 5 inch wide strips coming from hogs over 350 pounds to be ground into sausage. You'll often find this cut in oriental restaurants. You'll need to cull about 20 % due to scraping during processing. Ribs that have the bone showing a little meat should be discarded. Also, because these come from larger (tougher) hogs, they will need to stay in the wrapped stage longer (see rib cooking section), but having said that, they are usually 30% to 50% of baby back prices.

Refer to the pork meat chart on the following pages for more detailed cuts of meat.

NOTES

Beef Cuts for BBQ

Any of the tougher cuts of beef that you might use in a pot roast can benefit from the low and slow barbecue methods. The two cuts that are most often seen are beef ribs and beef brisket.

Beef ribs can be smoked in a process similar to pork ribs. Size-wise, the beef ribs will be much larger than spare ribs or a St. Louis cut. Beef ribs will also have a greater amount of fat. As the fat renders from the ribs you will have significant shrinkage that will expose the ends of the rib bones. As with pork spare ribs, you will find beef ribs easier to handle if you cut them into sections of 4 bones each. Beef ribs hold up well to mesquite smoke and stronger dry rubs. For your friends that do not eat pork, beef ribs are a good alternative. They are also a good candidate for a mustard pack that can eliminate the basting process since they contain so much fat.

Beef brisket is probably the toughest cut of beef you will encounter - tough in texture and tough to properly cook. Briskets need to be cooked from 8 to 12 hours or more for the proper smoke flavor and tenderness. The beef brisket is a lean, tough muscle, but it is also covered in a significant layer of fat. A whole brisket will weigh-in in the 12 to 15 pound range. For your first attempt, I would suggest buying a half-brisket. The brisket cut will be fairly thick at one end (the cap) and thinner at the trailing end (the tip). Because of this, the smaller tip often burns or "crisps up" before the thicker portion is ready. In some areas, these burnt ends are considered a delicacy. I consider them a "chef's reward" to be devoured while the pitmaster slaves over the rest of the brisket. I like to trim the fat cap down to ½ inch in thickness. The brisket will also benefit from an overnight soak in your marinade of choice, either a wet concoction or a blanket of dry rub. I prefer to cook my briskets fat-side-down. As with pork shoulders, I want to build layers of flavor and barq through the basting process. When the brisket is wrapped for the final stages or while cooling, I will place the meat fat-side-up to promote self-basting as the fat continues to render from the meat. Brisket should be brought to an internal temperature of 180° to 190° to reach the tenderness desired. Some cooks like to shred the brisket at that point for sandwiches. I prefer to let the brisket cool to around 150° so that it can be sliced in half-inch slices for serving or judging. Brisket is also great when you pull it from the refrigerator the next day and slice it deli-thin for sandwiches with spicy mustard while the brisket is still cold.

For a St. Patrick's Day surprise, buy a corned-beef brisket and smoke it like any other brisket. The cooking time will be greatly reduced since the cuts available in grocery stores are usually in the 5 to 6 pound size, but the combination of the cured brisket and hickory smoke will wow your friends. Brisket that has been brined for the 7 – 8 days necessary for corned-beef tends to be on the tough side. You will need to leave it in the wrapped stage long enough to become tender. Don't rush it! As long as you keep your temperature below 350° during the wrapped stage, the meat and juices will not burn. Your patience will be rewarded.

PORK CHART

RETAIL CUTS OF PORK — WHERE THEY COME FROM AND HOW TO COOK THEM

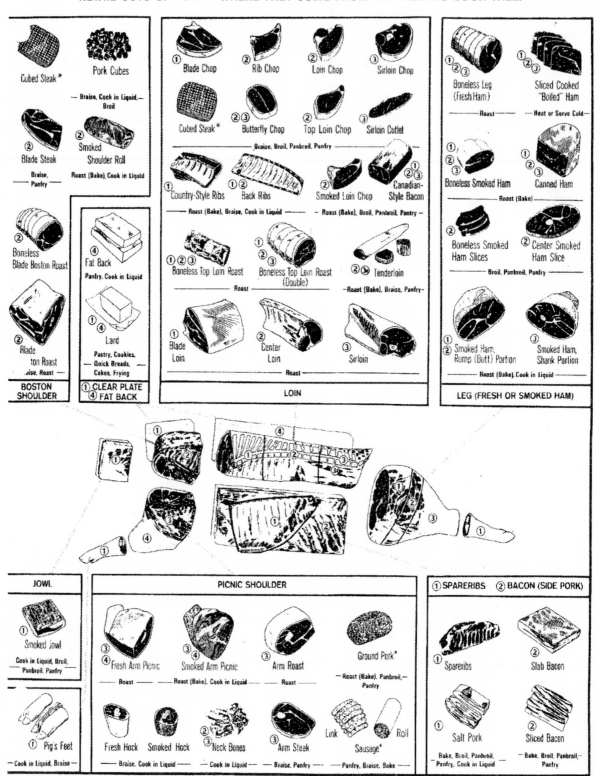

Cubed Steak* **Pork Cubes**

— Braise, Cook in Liquid, Broil

② **Blade Steak**

— Braise, Panfry —

② **Smoked Shoulder Roll**

Roast (Bake), Cook in Liquid

② **Boneless Blade Boston Roast**

② **Blade ton Roast**

...ise, Roast —

BOSTON SHOULDER

④ **Fat Back**

Pantry, Cook in Liquid

① ④ **Lard**

Pastry, Cookies, Quick Breads, Cakes, Frying

① **CLEAR PLATE**
④ **FAT BACK**

① **Blade Chop** ② **Rib Chop** ② **Loin Chop** ③ **Sirloin Chop**

②③ **Cubed Steak*** **Butterfly Chop** ② **Top Loin Chop** ③ **Sirloin Cutlet**

— Braise, Broil, Panbroil, Panfry —

① **Country-Style Ribs** ①② **Back Ribs** ② **Smoked Loin Chop** ①③ **Canadian-Style Bacon**

— Roast (Bake), Braise, Cook in Liquid —   — Roast (Bake), Broil, Panbroil, Pantry —

①②③ **Boneless Top Loin Roast** ①②③ **Boneless Top Loin Roast (Double)** ②③ **Tenderloin**

— Roast —   — Roast (Bake), Braise, Panfry —

① **Blade Loin** ② **Center Loin** ③ **Sirloin**

— Roast —

LOIN

①②③ **Boneless Leg (Fresh Ham)** ①②③ **Sliced Cooked "Boiled" Ham**

— Roast —   — Heat or Serve Cold —

①②③ **Boneless Smoked Ham** ①②③ **Canned Ham**

— Roast (Bake) —

② **Boneless Smoked Ham Slices** ② **Center Smoked Ham Slice**

— Broil, Panbroil, Pantry —

①② **Smoked Ham, Rump (Butt) Portion** ③ **Smoked Ham, Shank Portion**

— Roast (Bake), Cook in Liquid —

LEG (FRESH OR SMOKED HAM)

JOWL

① **Smoked Jowl**

Cook in Liquid, Broil, Panbroil, Pantry

① **Pig's Feet**

— Cook in Liquid, Braise —

PICNIC SHOULDER

④ **Fresh Arm Picnic** ④ **Smoked Arm Picnic** ③ **Arm Roast** **Ground Pork***

— Roast —   — Roast (Bake), Cook in Liquid —   — Roast —   — Roast (Bake), Panbroil, Pantry —

Fresh Hock **Smoked Hock** ② **Neck Bones** ③ **Arm Steak** **Link** **Roll** **Sausage***

— Braise, Cook in Liquid — — Cook in Liquid — — Braise, Panfry — — Panfry, Braise, Bake —

① **SPARERIBS** ② **BACON (SIDE PORK)**

① **Spareribs** ② **Slab Bacon**

① **Salt Pork** ② **Sliced Bacon**

Bake, Broil, Panbroil, Pantry, Cook in Liquid   — Bake, Broil, Panbroil, Pantry

BEEF CHART

RETAIL CUTS OF BEEF — WHERE THEY COME FROM AND HOW TO COOK THEM

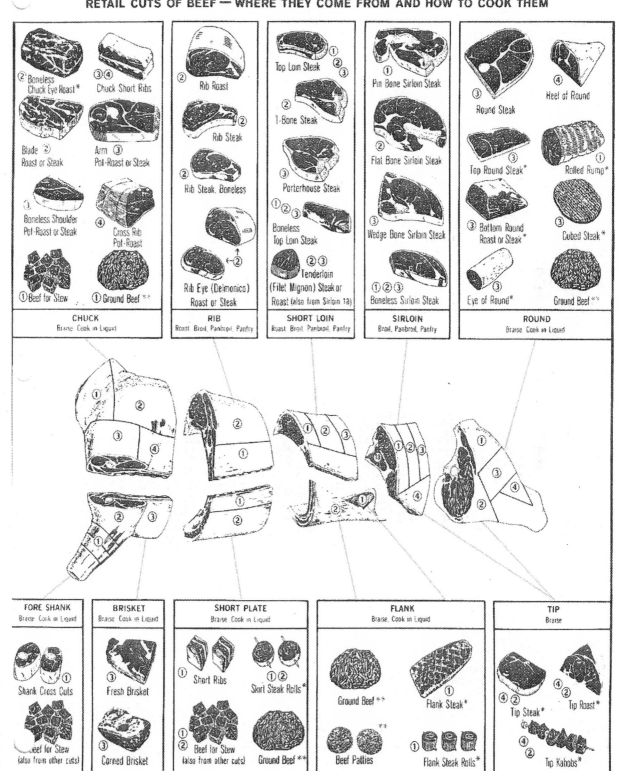

CHUCK
Braise, Cook in Liquid

- ② Boneless Chuck Eye Roast *
- ③④ Chuck Short Ribs
- Blade ② Roast or Steak
- Arm ③ Pot-Roast or Steak
- ③ Boneless Shoulder Pot-Roast or Steak
- ④ Cross Rib Pot-Roast
- ① Beef for Stew
- ① Ground Beef **

RIB
Roast, Broil, Panbroil, Panfry

- ② Rib Roast
- ② Rib Steak
- ② Rib Steak, Boneless
- ② Rib Eye (Delmonico) Roast or Steak

SHORT LOIN
Roast, Broil, Panbroil, Panfry

- Top Loin Steak ①②③
- T-Bone Steak ②
- Porterhouse Steak ③
- ①②③ Boneless Top Loin Steak
- ②③ Tenderloin (Filet Mignon) Steak or Roast (also from Sirloin 1a)

SIRLOIN
Broil, Panbroil, Panfry

- ① Pin Bone Sirloin Steak ②③
- Flat Bone Sirloin Steak ②
- ③ Wedge Bone Sirloin Steak
- ①②③ Boneless Sirloin Steak

ROUND
Braise, Cook in Liquid

- ③ Round Steak
- ④ Heel of Round
- ③ Top Round Steak *
- ① Rolled Rump *
- ③ Bottom Round Roast or Steak *
- ③ Cubed Steak *
- ③ Eye of Round *
- Ground Beef **

FORE SHANK
Braise, Cook in Liquid

- ① Shank Cross Cuts
- Beef for Stew (also from other cuts)

BRISKET
Braise, Cook in Liquid

- Fresh Brisket
- ③ Corned Brisket

SHORT PLATE
Braise, Cook in Liquid

- ① Short Ribs
- ①② Skirt Steak Rolls *
- ①② Beef for Stew (also from other cuts)
- Ground Beef **

FLANK
Braise, Cook in Liquid

- Ground Beef **
- ① Flank Steak *
- Beef Patties **
- ① Flank Steak Rolls *

TIP
Braise

- ④② Tip Steak *
- ④② Tip Roast *
- ④② Tip Kabobs *

39

Brining and Marinades

Brining is a very popular method of adding moisture and flavors to smoked turkeys and chickens. A basic brine is made of water and salt; with enough water to totally submerge the turkey overnight. The salt has a curing effect of destroying bacteria while adding flavor. Herbs and spices added to the brine can also impart additional flavors.

Brining for pork and beef is usually seen as a curing process and not so much as adding moisture. Most large cuts of pork and beef have sufficient fats and connective tissue that melt during the cooking process so that brining has minimal additional benefit. Brining only penetrates the meat to a depth of ½ inch. The addition of moisture to the surface does reduce the amount of moisture pulled from the inner area of the meat during the cooking process. The resulting moisture in the finished product may well be worth the effort.

For turkey brining or large cuts of meat you need a container large enough to accommodate your meat and several gallons of brine. If you meat floats to the top, you will need to weight it down with a plate or pot lid. During the brining process, the pot should be chilled in the refrigerator, an ice chest, or the back porch if it's Thanksgiving and cold enough outside (40 degrees or less).

Injecting of brines and marinades has proven to add moisture and flavors to meat products. Injecting can work very well, or it can be disastrous. The biggest mistake made is trying to inject too much liquid into one spot in your meat. In that case, the liquid simply pools up and does not disperse. Secondly, injecting baste with a high degree of solid particles (herbs and spices) can result in pockets of intense (bitter) flavors. To work well, injecting large numbers of small pockets is recommended.

For those adventurous souls, brining of a shoulder or ham can be done via the built-in artery system. Using the brachial or femoral artery, you can inject your solution through the existing circulatory system. This requires a 40 psi pressure to be effective. In most cases, the solution simply blows back through the artery due to insufficient pressure from most hypodermic-style injectors.

A process of injecting that works well for a competition team that is a consistent winner goes like this for shoulders, Boston butts and whole hogs:

Keep it simple. Don't try to impart too many flavors to your pork. Use a fruit juice, such as apple. Inject your meat before cooking using a large number of injection sites to maximize the distribution of juices. When you get to that midway point when the connective tissue and collagen are beginning to break down, inject again. At this time the meat will more readily accept the juice and the connective tissues will not restrict the dispersion of fluids. Lastly, inject again at the end of the cooking process when the muscle fibers have began to break down and you are lowering the heat. You should have minimal fluid loss going forward and dispersion of fluids should be uninhibited.

BBQ Marinades

Marinades can be seen as a less aggressive mode of brining. The goals are the same – adding flavor and moisture, but full immersion is not involved. Remember, even an overnight submersion will not achieve deep penetration, so marinades are more effective in adding flavors to the surface of meats.

A basic marinade can be as simple as a fruit juice or quick vinaigrette composed of one part vinegar to three parts oil. Start with the 1/3 acid (vinegar) to oil ratio and then add your desired flavors:

            Hot ………… chilies, peppers, hot sauce
            Sweet………..honey, molasses, ketchup
            Aromatic…….fresh or dried herbs, spices
            Light touch…..fruit juices or zest

Cooking teams will often use their basting sauce to marinate their meats from one hour to overnight. Plastic bags or plastic wrap can be used to protect the meat and hold your baste close to the surface during the process. To impart the maximum flavor you can apply dry rub to your meat and allow the rub to work itself into the meat before placing the meat in a bag with marinade. For an overnight process, you should turn the meat every 4 – 6 hours to make sure you get complete coverage.

Marinades are generally discarded after use because of the contact with raw meat. If you choose to use it as part of you basting process, you should boil the reserved liquids for five to ten minutes before reuse, or longer if you want to reduce it down and intensify the flavors.

Some general guidelines on marinade times:

Veggies	10-15 minutes	just enough to coat
Fish	1 hour max	acids may actually cook fish
Roasts/Chicken	4 – 8 hours	can go overnight
Ribs/Chops	8 – 12 hours	can go overnight

Dry Marinades (Rubs)

"It's not the size of your smoker it's how you rub your meat," said Reid Jackson, captain of the Holy Smokers Too Mile High BBQ team at the Dillon Colorado BBQ contest in 2005. A profound statement indeed!

There's probably no greater area of secrecy and pride than the area of dry rubs used in BBQ cooking. While the basic ingredients may start with similar herbs and spices, they only serve as the launch pad for the imaginations of BBQ cooks everywhere.

You can apply your rubs just before cooking or allow the spices to marinate on the meat overnight. Or, in true Memphis style vies-a-vie the Rendezvous Restaurant, you can sprinkle your rub on your meat right before serving.

The key ingredients of a typical rub include peppers, salt, sugar, garlic powder and mustard powder. Throw in some celery seed, oregano, or basil and you're on your way to developing your signature rub.

I prefer to omit salt from my rubs, unless I'm using it as a serving rub where salt can enhance the taste buds. There are two reasons: (1) I get enough salt in my diet on a daily basis and the peppers I use are a great salt-flavor substitute, (2) salt on raw meat will aggressively pull liquids from the meat. Remember, salt has been used for years to dehydrate country hams and preserve meat and fish.

Rubbing you meat is more a benefit for the chef and his (or her) theatrical performance for guests and onlookers. It is important that you get thorough coverage, but you will not really rub the spices deep into the meat. A light spray of apple juice will aid the adherence of your dry rubs. If you dry your meat prior to applying your rub, you will still have good luck. If you let your rub sit on the meat for any length of time you will notice that the surface will go from dry to pasty. The process of diffusion will pull moisture from the meat and pull the spices into the surface of the meat. Spraying the surface with juice or water will give you the same affect without encouraging fluids to be pulled from your meat.

Your dry rub is the essential first step in developing that BBQ barq (bark) that is relished by BBQ connoisseurs. In combination with meat proteins and your basting sauces, a crusty caramelized new skin will begin to develop on the surface of your meat. Throughout the cooking process you want to develop layers of flavor to enhance the smoked flavor of your meat. Fish and chicken do best with mild rubs. Pork does well on the sweet and spicy side. Robust beef, venison, lamb and wild duck can stand up to more pronounced, hearty herbs.

To protect you rubs and their essential oils responsible for much of the aroma and flavor, you should store you rubs in tightly sealed containers in a cool, dry, dark space. If you are using a flip-top shaker for storage, remove the top and place a piece of plastic wrap over the container and under the lid. Heat and air will dry out the essential oils. Moisture penetrating your container will cause caking. Rubs stored over time should be sifted before use to remove clots and aerate the spices to improve their "dusting" properties. Refrigeration will keep the spices dry and preserve the color of spices like chili powder and paprika. If you haven't used your rubs since last summer, it makes sense to make up a new batch for the upcoming season. Bland spices make for a bland finished product. While whole spices can keep from 2 – 5 years, ground spices begin to lose flavor after six months and rarely last longer than 2 years.

The Mustard Monster

Chuck Slaten from Hot Springs, Arkansas was given the moniker of "The Mustard Monster" when he was competing at Memphis in May in the 1980's. He won the rib category in 1986. He was famous for coating his ribs, after application of dry rubs, with mustard prior to placing his ribs on the smoker. Chuck moved to Texas and hasn't been seen much lately. During our *Decade of Dominance,* Holy Smokers Too also used a mustard pack on our ribs.

Basically you prepare your ribs or brisket as usual with your chosen dry rubs then, you coat the meat with a mustard pack of French's yellow mustard. Beef brisket does well with a coarse ground mustard or Dijon. The mustard pack seals in the meats juices, but adds little mustard flavor. Over the smoking process the mustard will dissipate such that you will not see any mustard on the surface of your ribs. After all, mustard is no more than BBQ spices in a vinegar base that evaporates leaving behind the spices. Since your ribs are coated with the mustard pack, basting is not required. This comes in very handy if you are trying to watch a football game or just wish to spend time with your guests away from the smoker. *If you don't want to get roped into kitchen duty you may not want to tell your spouse the ribs don't need your full attention.* You can follow your usual cooking routine. The ribs are ready when you no longer see the mustard on their surface and the meat begins to pull back from the bone. Raise one end of the rib rack and test for flexibility. The mustard will mix with your dry rub to form the new barq on the surface of your meat.

For the Blues, Brews & BBQ contest in Charlotte, North Carolina I taught my BBQ cooking class on Friday afternoon before the cook's meeting. During class I used a rack of St. Louis cut spare ribs to demonstrate the mustard pack method. Those ribs were placed in the smoker around 4:00 P.M. At the end of class, I pulled my smoker over to the backyard division to let J.J. Butz use the smoker to cook BBQ for friends and for their competition the next day. Around 11:00 P.M., as they were cleaning out the BBQ to make room for their competition meats, someone found the rack of mustard packed ribs in the rear of the smoker. They had been subjected to some seven hours of smoke and various temperatures. No one had basted the ribs because they were in the back corner of the smoker and ignored. We thought surely they would be 'crispy critters" and suitable for the dog bowl. To our surprise, they were moist and quite good – we fought over each rib! Now that's a real life testimony to the legacy of the Mustard Monster!

Marinade Recipes

Basic Brine

1 cup	coarse salt		½ cup	vinegar
1 cup	sugar		1 tablespoon	peppercorns
1 gallon	water		1 teaspoon	garlic powder

Combine all ingredients and stir until salt and sugar are dissolved. Place your meat in a container large enough to totally submerge your turkey or roast. Weight the meat down with a plate or pot lid if floating is a problem. Brine the meat overnight making sure the pot stays cool in the fridge, ice chest, or back porch if overnight temps are 40° or less. Remove the meat and pat dry before applying your dry rub prior to cooking.

Billy Bob Billy's
Basic Baste & Marinade

1 quart	Kraft Original BBQ Sauce
1 quart	apple cider
1 cup	Wickers marinade
1 cup	lemon juice

Combine all liquid ingredients, stirring in the BBQ sauce until dissolved. This is also great as an overnight marinade for chicken or ducks. Thin enough that it should not burn when used as baste, unless you are grilling at 400 degrees. The acidity of the lemon juice, apple cider, and Wickers works to tenderize the meat while adding flavor and moisture. Substitute your own BBQ sauce of choice.

Wickers marinade is distributed in the south at major grocers, but you can order your own from www.wickersbbq.com.

Citrus Adobo Marinade

½ cup	extra virgin olive oil – EVOO		1 cup	fresh lime juice
2	dried red chilies		1 tablespoon	adobo sauce
¼ cup	onion, minced		1	adobo pepper
2 cloves	garlic, minced		1 teaspoon	ground black pepper
1 teaspoon	cumin seeds		1 teaspoon	coarse salt
1 cup	orange juice			

Toast the cumin seeds on medium heat in a dry skillet or pan. Coarsely chop the adobo pepper. Add one tablespoon of olive oil and sauté the chilies and onion until the onion is translucent (five - ten minutes). Combine all the ingredients and allow them to cool slightly. Lightly pulse marinade in a food processor. Best used as soon as prepared.

Chicken can marinate overnight. White fish marinated for one hour and then dusted with a dry rub and grilled will light up your eyes.

Citrus Cilantro Marinade

½ cup	lime juice		½ teaspoon	cayenne pepper
½ cup	orange juice		½ teaspoon	coarse black pepper
3 tablespoons	lemon juice		¼ teaspoon	kosher salt
2 teaspoons	lemon zest		¼ teaspoon	oregano
2 teaspoons	lime zest		¼ teaspoon	paprika
½ cup	extra virgin olive oil		¼ teaspoon	oregano
2/3 cup	chopped cilantro			
4 cloves	garlic, pressed			

- Tabasco, habanero, or jalapeno to taste.
- Yields two cups

Combine three juices and zest, stirring to mix. Drizzle in olive oil while stirring. Add spices, cilantro and pressed garlic for final stirring. (Can be made in food processor). Allow marinade flavors to marry for two hours in refrigerator before using.

Especially good on chicken and firm white fish. Fish should only marinade for thirty minutes to prevent the acids from pre-cooking the fish. Chicken can marinate overnight.

Jamaican Jerk Marinade

2 cups	green onions	2 teaspoons	fresh ginger
½ cup	Vidalia onion	1 teaspoon	ground allspice
2 tablespoon	white vinegar	¼ teaspoon	ground nutmeg
1 tablespoon	soy sauce	¼ teaspoon	black pepper
1 tablespoon	vegetable oil	¼ teaspoon	ground cinnamon
2 teaspoon	coarse salt	2 cloves	garlic, minced
2 teaspoons	fresh thyme	2	habanero peppers
2 teaspoons	brown sugar		

Coarsely chop 2 cups of green onions and ½ cup of Vidalia onion, as well as habanero peppers and ginger. *You will want to remove the seeds and membranes inside the peppers prior to chopping.* Place ingredients in a blender and slightly pulse until well mixed. Marinade can be used to marinate pork, chicken or brisket for one hour or overnight.

For large cuts of meat, a batch should be made for basting midway in the cooking process, or for a real kick, in the last hour to thirty minutes.

Billy Bob Billy's
Lamb Marinade

½ onion	medium sized	1 teaspoon	lemon zest
1 head	fresh garlic	2 teaspoon	sea salt
1" cube	fresh ginger	2 teaspoon	coarse black pepper
2 tablespoons	sweet paprika	6 tablespoons	olive oil
1 tablespoon	cumin seeds		

Rough chop the onion and grind cumin seeds in spice grinder or with mortar and pestle. Pulse dry ingredients in food processor or blender and slowly drizzle in the olive oil to make a loose paste. Apply mixture generously to lamb, place in plastic bag and marinate overnight.

Margarita Paste

¼ cup	lime juice	2 tablespoons	kosher salt
½ bunch	cilantro	1 tablespoon	black peppercorns
2	jalapeno peppers	1 teaspoon	lime zest
2	cloves garlic	1 teaspoon	yellow mustard
2 tablespoons	extra virgin olive oil	1 teaspoon	brown sugar

Rough chop cilantro. Seed and chop two jalapenos. Rough chop two cloves of garlic. Zest one whole lime. Crack peppercorns. Combine wet ingredients (lime juice, mustard, EVOO in a separate bowl. Combine all dry ingredients (except cilantro) and mix well. Drizzle wet ingredients into dry while lightly tossing. Add cilantro and mix until a consistent paste is formed. Coat the meat to be cooked approximately 30 minutes before placing on the grill or in the smoker.

Works well with chicken and mild fish.

Billy Bob Billy's
Bloody Mary Marinade

1 cup	Clamato juice	2 tablespoons	kosher salt
1 cup	tomato juice	1 tablespoon	black peppercorns (cracked)
¼ cup	lime juice		
½ bunch	cilantro	1 teaspoon	lime zest
2	jalapeno peppers	1 teaspoon	BBQ dry rub
2	cloves garlic		

Rough chop cilantro. Seed and chop two jalapenos. Rough chop two cloves of garlic. Zest one whole lime. Combine wet ingredients in a separate bowl. Combine all dry ingredients (except cilantro) and mix well. Drizzle wet ingredients into dry while lightly tossing. Add cilantro and mix well. Marinade at least one hour up to 24 hours.

Great with robust cuts of beef and venison.

Habanero – Garlic Vinaigrette

6 cloves	roasted garlic	¾ cup	EVOO
½ medium	habanero pepper		(extra virgin olive oil)
¼ cup	lime juice	2 tablespoons	honey
¼ cup	chopped cilantro	2 tablespoons	water
Salt & Pepper to taste			

Take one whole head of garlic and slice ½ inch of the top off to expose the cloves. Sprinkle with olive oil, salt and pepper and wrap in aluminum foil. Place on hottest spot in your smoker for 45 minutes to one hour (30 min. in 350° oven). Unwrap garlic and squeeze six cloves into mixing bowl. Mash garlic with fork into a paste. Halve one medium habanero pepper and remove seeds and membrane. Mince pepper. Combine all ingredients in bowl and whisk until smooth and creamy. Add pinch of salt and ground pepper to your taste.

Great with pork, chicken or fish.

Smokey Vinaigrette

6	plum tomatoes	½ tablespoon	Adobo sauce
½ cup	olive oil	1 clove	garlic, chopped
½ cup	red wine vinegar	2 tablespoon	basil leaves, chopped
Salt and pepper to taste			

Roast tomatoes on grill until skins are blackened. Place tomatoes in a paper bag and allow them to steam for 15 minutes. Remove skins and coarsely chop the plum tomatoes. Combine first four ingredients in a food processor and pulse until mixed well. Continue pulsing and add garlic, basil, salt and pepper.

Works well as baste with chicken or other fowl. Can also be used as an overnight marinade.

Baste Recipes

Holy Smokers Too
Rib Baste

108 oz.	BBQ Sauce
108 oz.	Apple Cider
28 oz.	Wickers Baste
28 oz.	Lemon Juice

Combine liquids in a large bowl or jug. Drizzle in BBQ sauce while stirring to dissolve sauce into the liquids. Baste keeps well in the refrigerator for 30 days.

Wickers marinade is distributed in the south at major grocers, but you can order your own from www.wickersbbq.com.

Whole Hog Baste

1 gallon	Apple Cider Vinegar
1 quart	Worcestershire Sauce
1 1/4 cup	Lemon Juice
3 tablespoons	Black Pepper

Combine all ingredients and bring to a hard boil 10-15 minutes. Smoke hog, belly down, for six hours. Once turned, baste hog on the hour until the last hour when you should aspirate the body cavity and remove the liquids that have pooled over the ribs. Shoulders and hams should measure 185°. Allow cavity to dry out during the last hour. A simple baste with historic roots.

In the early years, North Carolina hog cookers used a simple baste of vinegar and black pepper because that is what was available. That tradition continues today in many North Carolina restaurants, pits and BBQ contests.

REDNECK BAR B Q EXPRESS

Basting Sauce

2 gal	white vinegar	2 oz	allspice
8 oz	Liquid Smoke	1 oz	onion powder
2 lbs	brown sugar	2 oz	garlic powder
10 oz	salt	2 oz	chili powder
3 oz	black pepper		

Combine all ingredients and bring to a boil.
Reduce heat and simmer 3 hours.

Hogaholic's Baste

4 cups	Wickers Marinade and Baste
2 cups	apple cider vinegar
2 cups	vegetable oil
1/2 cup	lemon juice

Combine all ingredients and whisk until well incorporated.
Wickers marinade is distributed in the south at major grocers, but you can order your own from www.wickersbbq.com.

Smokin' Friars
Mustard-Vinegar Baste

3 tablespoons	olive oil	2 teaspoons	mustard seeds
1 medium	Spanish onion	2 teaspoons	coriander seeds
2 cloves	garlic	¼ cup Dijon	mustard
1-1/2 cups	cider vinegar	2 tablespoons	dry mustard
½ cup	water	1 tablespoon	Worcestershire sauce
2 tablespoons	sugar		

Heat the oil in a medium saucepan over medium heat. Finely chop onion and garlic. Add onion and garlic and cook, stirring occasionally, until soft, about 3-4 minutes. Add the vinegar, water, sugar, mustard and coriander seeds and bring to a simmer. Cook for 5 minutes, then remove from the heat. Whisk in the Dijon, dry mustard and Worcestershire. The sauce can be made up to 1 day in advance. Reheat before using as baste.

Good on pork, chicken and beef brisket.

Dry Rub Recipes

Billy Bob Billy's
Basic Rub

2 tablespoons	sweet basil	1 tablespoons	garlic powder
2 tablespoons	coarse black pepper	2 teaspoons	celery seed
2 tablespoons	paprika	1 teaspoon	dry mustard
1 tablespoons	Chile powder	2 teaspoons	ground cumin
½ tablespoon	red pepper – Cayenne		

Billy Bob Billy's
Finishing Rub

2 tablespoons	sweet basil	1 tablespoon	kosher salt
2 tablespoons	coarse black pepper	2 teaspoons	celery seed
2 tablespoons	paprika	1 teaspoon	dry mustard
1 tablespoons	Chile powder	2 teaspoons	ground cumin
½ tablespoon	red pepper – Cayenne	4 tablespoons	brown sugar
1 tablespoons	garlic powder		

Whisk the ingredients for these two rubs and store in a cool, dry space. If you notice caking of your rub before use (none of us really has space in the fridge or spice cellar to store our rubs) run your rub through a sifter to loosen-up or aerate the spices. If it's been a year since you last used these rubs – throw them away and start over. You can't have a great product if you start with less-than-great spices. You may want to add the brown sugar to the finishing rub (the only difference) right before applying, since it tends to exacerbate caking. For those of you that insist on salt to awaken your taste buds, you can add one tablespoon of coarse salt to the finishing rub. As previously stated, I don't use salt in my basic rub because I don't want to pull moisture from my products.

For ribs, I like to hold the basil back until the other spices are applied and then add a coat of basil for the visual effect. Also, the basil falling off into the fire always gives some competing team members a 60's flashback as the aroma drifts their way.

Billy Bob Billy's
Best Brisket Rub

This rub takes a time to prepare, but the fresh-ground spices and the extra effort are well worth the results.

2 dried	chipotle peppers	¼ cup	sweet paprika
2 tablespoons	black peppercorns	1 tablespoon	garlic powder
1 tablespoon	cumin seeds	1 tablespoon	sweet basil
1 tablespoon	coriander seeds	1 tablespoon	onion powder
1 tablespoon	mustard seeds		

Toast seeds in a dry cast iron skillet over medium heat (2- 4 min). Grind seeds, peppers and peppercorns with mortar and pestle into coarse powder. *You can use your spice grinder if you have one.* Combine all ingredients and apply liberally to your brisket. The Rub can be applied right before cooking, but will have a greater impact if allowed to marinate the brisket overnight.

Billy Bob Billy's
Lamb Rub

¼ cup	sea salt	1 tablespoon	rosemary
¼ cup sweet	paprika	1 tablespoon	marjoram
1 tablespoon	black peppercorns (cracked)	1 tablespoon	lavender
		1 teaspoon	cumin seeds, ground
1 tablespoon	garlic powder		
1 tablespoon	thyme		

Finely chopped fresh thyme, rosemary, and lavender will give you great bursts of flavor from this rub. It is extra effort, but freshly grinding the cumin and pepper will also bring great rewards. Apply liberally and allow meat to marinate from 1 hour to overnight.

Billy Bob Billy's
Beef Tenderloin Rub

2 tablespoons	sea salt	2 tablespoons	rosemary
2 tablespoons	black pepper corns	2 tablespoons	lavender
2 tablespoons	sweet basil	1 teaspoon	cumin seeds
2 tablespoons	thyme	1 teaspoon	coriander seeds

Prepare tenderloin for grilling. Rub beef with olive oil. Finely chopped fresh basil, thyme, rosemary, and lavender will give you great bursts of flavor from this rub. It is extra effort, but freshly grinding the cumin, coriander and pepper will also bring great rewards. After all, this is the king of beef cuts. Crack peppercorns, grind cumin and coriander seeds and mix ingredients. Sprinkle with rub or sprinkle rub on cookie sheet and roll tenderloin until coated with rub.

Cajun Blackening Spices

5 teaspoons	paprika	1/2 teaspoon	black pepper
1 teaspoon	ground dried oregano	1/2 teaspoon	white pepper
1 teaspoon	ground dried thyme	1/2 teaspoon	garlic powder
1 teaspoon	cayenne pepper		

Mix together and store in an air-tight container.
Great on fish for a quick smoke and intense flavors when grilling or pan searing.

Carolina BBQ Rub

2 tablespoons	salt	2 tablespoons	black pepper
2 tablespoons	sugar	1 tablespoon	cayenne pepper
2 tablespoons	brown sugar	1/4 cup	paprika
2 tablespoons	freshly ground cumin		
2 tablespoons	chili powder		

Combine all ingredients in a small bowl and mix well.
Great as a dry rub on beef, chicken, lamb or pork.

Chili Paste

1	Lemon (and zest)	1 tablespoon	olive oil
1	Lime (and zest)	1 tablespoon	paprika
1	orange (and zest)	1 teaspoon	ground cumin
1/2	green chili, chopped	1 teaspoon	salt
5 cloves	garlic (finely crushed)	1/2 teaspoon	dried oregano
3 tablespoons	mild chili powder	1/4 teaspoon	ground cinnamon

Mix ½ teaspoon of each fruit zest and all the juices with the other ingredients. Add more spices if desired. Let stand at least 30 minutes before using (should thicken).
Great on chicken and fish.

Espresso Rub

¼ cup	ground espresso beans	1 tablespoon	ground black pepper
¼ cup	Ancho chili powder	1 tablespoon	ground cumin
2 tablespoons	Spanish paprika	1 tablespoon	dried oregano
2 tablespoons	dark brown sugar	2 teaspoons	ground ginger
1 tablespoon	dry mustard	2 teaspoons	cayenne
2 teaspoons	kosher salt		

Mix all ingredients and apply liberally to pork and beef. This is a flavorful rub and can be applied just before smoking or left on overnight. This rub comes from the Smokin' Friars and is used on their Espresso Ribs. It is also good on brisket and wild game.

Hogaholics Dry Rub

2 tablespoons	coarse salt	1 tablespoon	chili powder
2 tablespoons	sugar	1 tablespoon	paprika
1 tablespoon	lemon zest	½ tablespoon	black pepper
1 tablespoon	garlic powder	½ tablespoon	white pepper
1 tablespoon	onion powder	½ tablespoon	cayenne pepper

The Hogaholics used this rub on their ribs and shoulders. The lemon zest adds a nice kick. Used on both ribs and pork shoulders.

Jack's Old South Rub

¼ cup	brown sugar	2 teaspoons	garlic powder
¼ cup	sweet paprika	2 teaspoons	onion powder
¼ cup	kosher salt	1 teaspoon	cayenne pepper
3 tablespoons	black pepper	1 teaspoon	dried basil

This is a published version of their rub. Only Myron Mixon knows for sure and he's not talking. You can't go wrong with a consistent winner.

It's good on all cuts of pork and chicken.

Jamaican Jerk Rub

4	Habanero peppers	2 cloves	garlic, minced
2 teaspoons	sea salt	1 teaspoon	ground allspice
2 teaspoons	brown sugar	¼ teaspoon	black pepper
2 teaspoons	fresh ginger	¼ teaspoon	ground nutmeg
2 teaspoons	thyme	1/8 teaspoon	ground cinnamon

This rub is not for the timid. Be sure and wear gloves when handling the habanero peppers. You can reduce the habanero peppers, or replace with jalapeno peppers, but then it wouldn't be jerk. Seed the habanero and slice micro-julienne style. These little slices of "carrot" will effuse heat into the rub as well as give you a micro-burst of heat when you bite into one.

Kansas City Rib Rub

1/2 cup	brown sugar	1 tablespoon	chili powder
1/4 cup	paprika	3/4 tablespoon	garlic powder
1 tablespoon	black pepper	3/4 tablespoon	onion powder
1 tablespoon	salt	1 teaspoon	cayenne

This looks a lot like Jack's Old South with a slight variation. Could be a good launch pad for your own favorite.

REDNECK BAR B Q EXPRESS

Dry Rub

6 tablespoons	celery salt		
6 tablespoons	granulated garlic	3 tablespoons	mustard powder
6 tablespoons	black pepper	3 tablespoons	cumin powder
6 tablespoons	paprika	3 tablespoons	Accent
6 tablespoons	lemon pepper	3 tablespoons	meat tenderizer
3 tablespoons	onion powder	3 tablespoons	chipotle pepper
3 tablespoons	chili powder	1 1/5 tablespoons	hickory salt

These guys are always one of Suzie Que's favorites at Memphis in May because of their friendliness and marinated strawberries. Their barbecue has always been quite tasty.

Billy Bob Billy's
Venison Rub

6 tablespoons	ground coffee	2 teaspoons	onion powder
2 tablespoons	sea salt	1 teaspoon	ground cumin
2 tablespoons	sweet paprika	1 teaspoon	coriander
2 tablespoons	black pepper	1 teaspoon	unsalted cocoa
2 tablespoons	sweet basil		powder.
2 teaspoons	garlic powder		

Toast coriander seeds in a dry skillet for 2 - 4 minutes. Crush with mortar and pestle into coarse powder. (Use back of wooden spoon in the skillet as substitute). Combine all ingredients in a bowl and whisk until uniform. Rub into venison and allow to marinate one hour (or overnight) before placing in smoker.

Billy Bob Billy's
Citrus Venison Rub

1 tablespoon	coriander seeds		2 teaspoons	black pepper
1 tablespoon	orange zest		1 teaspoon	coarse salt
1 tablespoon	lime zest		1 teaspoon	thyme

Toast coriander seeds in a dry skillet for 2 - 4 minutes. Crush with mortar and pestle into a coarse powder. (Use back of wooden spoon in the skillet as substitute). Combine all ingredients in a bowl and whisk until uniform. Rub into venison and allow one hour (or overnight) before placing in smoker. The lime zest really wakes up the spices.

BBQ Sauce Recipes

Billy Bob Billy's
Basic BBQ Sauce

2 cups	ketchup	2 tablespoons	molasses or honey
¼ cup	apple cider	2 tablespoons	yellow mustard
	(vinegar as option)	1 teaspoon	Tabasco sauce
¼ cup	Worcestershire	1 tablespoon	BBQ dry rub
¼ cup	brown sugar		

Bring ingredients to a boil and simmer for 15 – 20 minutes, stirring occasionally. This started out as my Mom's recipe and grew on me from there. It's simple, but works well on pork and chicken.

Beer Barbecue Sauce

1 cup	barbecue sauce	2 tablespoons	Dijon mustard
1 cup	ketchup	1 tablespoon	Worcestershire sauce
2/3 cup	dark beer	1 teaspoon	hot pepper sauce
1/4 cup	honey or molasses	1/2 teaspoon	pepper
2 tablespoons	lemon juice	2 cloves	garlic, minced
2 tablespoons	red wine vinegar	2	onions finely chopped

In large bowl, combine barbecue sauce, ketchup, beer, honey, lemon juice, vinegar, mustard, Worcestershire sauce, hot pepper sauce, pepper, garlic and onions. Place food in marinade and let stand at room temperature for up to 2 hours or in refrigerator overnight. When ready to cook, remove food and place marinade in saucepan, then cook for 10 minutes or until thickened. Spoon off fat as it rises to the surface. Sauce can be used for basting or a serving sauce after thickening.

Holy Smokers Too – BBQ Sauce

Liquid ingredients

1 stick	butter (not margarine)		3 oz	white vinegar
3	18 oz bottles		2 oz	"Louisiana Hot Sauce"
	"Kraft Original BBQ Sauce"		2 oz	Worcestershire sauce
3 oz	lemon Juice		1 oz	soy sauce

Spices

½ cup	brown sugar		1 ½ teaspoons	red pepper
2 teaspoons	dried mustard		1 teaspoon	thyme
1 ½ teaspoon	garlic powder		1 teaspoon	chili powder
1 teaspoon	ginger (powder)		1 teaspoon	cumin
1 ½ teaspoon	onion powder		1 teaspoon	allspice
1 ½ teaspoons	cinnamon		1 teaspoon	celery seed
1 teaspoon	oregano		1 teaspoon	dill weed
1 teaspoon	sweet basil		1 teaspoon	black pepper

Melt butter and then mix in all liquid ingredients. Add all spices, then mix with whisk. Cook on medium heat for 30-45 minutes ensuring that sauce is stirred often. Let cool for 30 minutes then refrigerate. Gets hotter with time. Makes ½ gallon.

Note: Holy Smokers Too original sauce recipe is known to have won a Blue Ribbon at the 1947 Mid South Fair. The recipe was the product of Rick Cooper's father. It's been prodded, poked, and "improved" over the years to such an extent that Rick is probably the only one who knows the real recipe. And, unless he has it written down somewhere, we have all killed too many brain cells to recall where we started. Our general policy is to make three different batches and vote as to which one to use on contest day. As you might guess, it makes it hard for competitors (or new team members) to copy our sauce.

Sly as a bunch of gray-haired foxes, ain't we!!

Jack Daniel's Grilling Sauce

1/2 cup	pineapple juice
3 tablespoons	soy sauce
1 1/2 teaspoons	garlic powder
1/4 cup	Jack Daniel's Whiskey

Combine all ingredients and mix well. Dip meat in sauce and place on grill over hot coals. When meat is turned, brush with sauce. Grill to desired degree of doneness. Just before meat is removed from grill, brush again with sauce. Great with chicken or pork chops.

Alabama White Sauce

1 cup	mayonnaise	1 teaspoon	sugar
¾ cup	white vinegar	1 teaspoon	horseradish sauce
1 tablespoon	lemon juice	1 teaspoon	salt
1 tablespoon	black pepper		

Combine all ingredients and refrigerate until use (due to the mayonnaise). A well known Alabama team, Big Bob Gibson Bar-B-Q, headed by Chris Lilly, will marinate their chicken in this sauce and then dredge the finished product through the sauce hot off the smoker and on the way to the table. The flavors optimize when allowed to get to know each other overnight in the fridge.

This is "a recipe" for white sauce. Every competition team will give you "a recipe" but none will give you "the recipe."

You can get "the recipe" in bottle form at www.bigbobgibson.com.

Hogaholics BBQ Sauce

46 oz	tomato juice	2 teaspoons	black pepper
24 oz	ketchup	2 teaspoons	dry mustard
10 oz	soy sauce	1 teaspoon	garlic powder
10 oz	Worcestershire sauce	1 teaspoon	onion powder
2 cups	cider vinegar	1 teaspoon	oregano
2	lemons juiced	1 teaspoon	allspice
2 cups	brown sugar	1 teaspoon	ginger
2 teaspoons	cayenne	1 teaspoon	basil

Mix all wet ingredients then stir in dry spices. Simmer the sauce for one hour. This can be used as a serving sauce or rib glaze prior to serving. Recipe makes one gallon.

Hog Wild Steak Sauce

2 cups	ketchup	2 tablespoons	maple syrup
¼ cup	prepared horseradish	3 tablespoons	Ancho chili powder
2 tablespoons	honey	Pinch of kosher salt	
2 tablespoons	Dijon mustard	Pinch butcher grind black pepper	

Combine ingredients and mix thoroughly. Great as a serving sauce with your steaks or lightly brushed on for the last minute of grilling. Makes approx. 25 servings

You can order your own at www.hogwildbbq.com.

Beer Barbecue Sauce

		2 tablespoons	Dijon mustard
1 cup	barbecue sauce	1 tablespoon	Worcestershire sauce
1 cup	ketchup	1 teaspoon	hot pepper sauce
2/3 cup	dark beer	1/2 teaspoon	black pepper
1/4 cup	honey or molasses	2 cloves	garlic, minced
2 tablespoons	lemon juice	2	onions finely chopped
2 tablespoons	red wine vinegar		

In a large bowl, combine barbecue sauce, ketchup, beer, honey, lemon juice, vinegar, mustard, Worcestershire sauce, hot pepper sauce, pepper, garlic and onions. Place food in marinade and let stand at room temperature for up to 2 hours or in refrigerator overnight. When you are ready to cook, remove food and place marinade in saucepan and cook for 10 minutes, or until thickened. Sauce can be used for basting or serving with cooked food.

Honey Spiced BBQ Sauce

1¼ cup	ketchup	2 tablespoons	dry mustard
2/3 cup	salad oil	3 teaspoons	ginger, fresh grated
3/4 cup	vinegar	1	lemon, sliced thinly
5 tablespoons	Worcestershire sauce	3 tablespoons	butter
1 cup	honey		

Combine all ingredients in a saucepan and heat to blend together.
Remove lemon slices before basting.

Hoppin' John's BBQ Sauce

1	cup	yellow mustard	1	tablespoon	onion powder
1	cup	Worcestershire	1	tablespoon	ground black pepper
1	cup	brown sugar	1	tablespoon	Chile powder
1	cup	honey	1	tablespoon	salt
¼	cup	apple cider vinegar	½	tablespoon	thyme
1	tablespoon	garlic powder	1	teaspoon	cayenne
3	cups	ketchup	½	teaspoon	cumin
		(he preferred catsup)			

Add all ingredients except sugar and honey and bring to a slow simmer for 15 minutes. Turn off heat and dissolve sugar and stir in the honey. Yields two quarts. Because of the sugar and honey, this sauce is recommended as a serving sauce or may be used to glaze your BBQ during the last 15 minutes of the cooking process. *Too much heat and it will blacken!!*

Molasses Orange Barbecue Sauce

1	can	tomato soup, condensed (10 oz)	1	tablespoon	seasoned salt
			1	tablespoon	dry mustard
1	can	tomato sauce, 8 oz	1	tablespoon	Worcestershire sauce
1/2	cup	molasses, light	1	tablespoon	orange zest
1/2	cup	vinegar	1 1/2	teaspoon	paprika
1/2	cup	brown sugar, packed	1/2	teaspoon	pepper, black
1/4	cup	vegetable oil	1/4	teaspoon	garlic powder
1	tablespoon	minced onion, instant			

In a saucepan, combine all ingredients. Bring to a boil; reduce heat and simmer, uncovered, for 20 minutes. Use to baste beef or poultry the last 15 minutes of grilling.

Kansas City Style Sauce

2 cups	ketchup	1 teaspoon	onion powder
1 cup	molasses	1 teaspoon	garlic powder
½ cup	white vinegar	½ teaspoon	allspice
1 teaspoon	Chile powder	½ teaspoon	cinnamon
1 teaspoon	paprika	½ teaspoon	black pepper

Add all ingredients except molasses and bring to a slow simmer for 15 minutes. Turn off heat and stir in the molasses. Because of the molasses, this sauce is recommended as a serving sauce or may be used to glaze your BBQ during the last 15 minutes of the cooking process. *Too much heat and it will blacken!!*

Mustard Sauce

1/3 cup	brown sugar	1/4 cup	mustard
1/4 cup	onion, finely chopped	1/2 teaspoon	celery seed
1/4 cup	vinegar	1/4 teaspoon	garlic powder

For sauce, in a saucepan combine brown sugar, onion, vinegar, mustard, celery seed and garlic powder. Bring to boiling, stirring till sugar dissolves.

North Carolina Vinegar Sauce

Basic NC Vinegar Sauce

1 cup	Whitehouse cider vinegar
1 cup	water
1 tablespoon	red pepper flakes

My first taste of North Carolina vinegar sauce was at the first Blues, Brews & BBQ contest in Charlotte, NC. My newest, best friend I met at the contest was J.J. Butz. Now, J.J. had two mason jars with clear liquids that he claimed were the secrets to his success on the BBQ circuit. One jar had red pepper flakes floating in it. The other jar had a peach in the bottom of the jar. Both of these elixirs took my breath away with each swig. I took a liking to the jar with the peach in it. The vinegar and pepper was too stark for my liking. As we finished off that peach at around two a.m. in the middle of the night, I gained an understanding as to why J.J. Butz has yet to win a contest.

> *"It's not true NC sauce if you don't use Whitehouse apple cider vinegar,"*
> *says J.J. Butz.*

A "flavorful" variation

2 cups	cider vinegar
½ cup	brown sugar
3 teaspoons	salt
1 teaspoon	black pepper
1 teaspoon	white pepper
½ teaspoon	cayenne

Heat either recipe for 15 minutes and decant into your favorite mason jar.

Please label the basic sauce, since it has little color and might be mistaken for other elixirs stored in mason jars. *A whole new rendition of "fire water."*

Young Butz
Some-day-award-winning BBQ Sauce

2 cups	brewed black coffee	½ cup	soy sauce
2 cups	Worcestershire sauce	1	Anaheim pepper
3 cups	ketchup	1 medium	onion
2 cups	apple cider vinegar	8 cloves	garlic
2 cups	brown sugar	1 tablespoon	cumin seeds, crushed
1 cup	honey	2 tablespoons	paprika
½ cup	Dijon mustard	5 tablespoons	chili powder

Juice of one lemon
Salt and pepper to taste
2 – 3 tablespoons of red pepper are optional

Fine chop the pepper, onion, and garlic. Combine all ingredients in a large stock pot, bring to a boil, lower heat and let simmer 2 to 4 hours. Strain into jars and it's ready to use.
Makes 8 cups of sauce.

"The longer this sauce sits, the better it gets," says Joey.

He also admits to re-introducing some of the spices into the sauce after straining and tinkering with the ingredients according to taste. It took me two years to get Joey to share this version of his sauce. These young guys are suspicious of their elders?!?

Suzie Que loved it on her smoked turkey leg in Charlotte.

NC BBQ Sauce with a history

Jim Early, founder of the North Carolina BBQ Society listed the following basic recipe in his detailed history of North Carolina BBQ titled "The Best Tar Heel Barbecue Manteo to Murphy."

2 quarts	apple cider vinegar
2 ounces	red pepper flakes
2 tablespoons	salt
2 tablespoons	black pepper

Billy Bob Billy's
NC BBQ Sauce

6	habanero peppers	1 teaspoon	salt
2 cups	yellow onion, chopped	2 tablespoons	tomato paste
2	garlic cloves	2 cups	apple cider vinegar
4	allspice berries, crushed	2 cups	water
½ teaspoon	ground cumin	(more water if you want a thinner sauce)	
1 teaspoon	sugar		

Remove seeds and membranes from habanero peppers. Combine peppers, onion, garlic, allspice, and cumin. *Fresh crushed allspice berries are best because you get more of the oil from the berries.* If not, measure approx. ½ tablespoon of crushed allspice. Grind dry ingredients in a food processor until the y are a fine mixture. Combine all ingredients in a non-reactive saucepan and bring to a boil. Allow the mixture to cool and then place in a sealed bottle and refrigerate until used. Because of the boiling process and vinegar content, this sauce will keep for a long period in the refrigerator.

For a great hot sauce to be used on a wide variety of foods, just eliminate the 2 cups of water.

Although my North Carolina friends may see this as an abomination of true NC BBQ sauce – I just had to add a little more flavor and kick to their basic recipes.

South Carolina Mustard Sauce

1 cup	yellow mustard	2 teaspoon	hot sauce
1 cup	cider vinegar	2 teaspoon	Worcestershire
½ cup	sugar	1 ½ teaspoon	black pepper
2 tablespoons	margarine	1 teaspoon	salt

Melt margarine (*hey, we're on holiday, go ahead and use butter*) over low heat. Add the reset of the ingredients during a slow simmer for 30 minutes stirring regularly.

REDNECK BAR B Q EXPRESS

Glazing Sauce

1.5 lg bottles	Kraft BBQ Sauce	3 tablespoons	ground chipotle
½ bottle	Worcestershire Sauce	1 tablespoon	red pepper
1 cup	apple cider vinegar	¼ bottle	Pickapeppa Sauce
1 jar	honey	1 tablespoon	liquid smoke
½ pound	brown sugar		(mesquite)
¾ bottle	Tiger sauce	1 tablespoon	liquid smoke
1 tablespoon	seasoned pepper		(hickory)

Suzie Que always visits these guys at Memphis in May. They serve (Ladies Only) marinated strawberries filled with Kahlua and whipped cream. I enjoy her eating them almost as much as she does. ***Gotta see it to believe it!***

Tennessee BBQ Sauce

1 cup	catsup	1/2 cup	brown sugar
1/2 cup	cider vinegar	1 teaspoon	brown mustard
1/4 cup	Worcestershire sauce	1 teaspoon	celery seed
2 cups	water	1/2 teaspoon	salt
1	medium onion		

Coarse chop the onion. Combine ingredients in small saucepan and bring to boil. Simmer until reduced to thick sauce, stirring occasionally. Simple, but tasty. For those of you from the North, catsup is the original term for the tomato based sauce now known as ketchup.

Cooking Whole Hog

Holy Smokers Too has always been a whole hog cooking team. We've won our share of competitions in shoulders and ribs, occasionally pulling a hat trick and winning all three categories in local contests. As our members have aged, we have limited competing in all three categories as well as limiting the number of contests we enter. The slow, deliberate process of smokin' a whole hog fits our team's temperament. And besides, that gives us more time to tell lies about past performance. Susie Que prefers to say we have embellished our history. Most members admit to suffering from 'Old Timers' disease.

Cooking whole hog is the ultimate smokin' challenge. Most competitors only cook one hog, so it's an all-or-nothing process that takes 24 hours of time, attention, and patience.

It reminds me of the time I was driving to Memphis in May for a contest and I spotted a pig farmer on the side of the road on I-40. He was struggling with a flat tire, so I pulled over and offered to help.

"Are you taking these hogs to Fineberg Packing in Memphis?" I asked.

"Nope!" he said. "I'm driving these hogs to Texas."

"Why Texas," I asked.

"I can get two cents a pound more in Texas," he said.

"That doesn't make much sense," I said, "because it will take two days to drive there and two days back. Seems like and awful waste of time," I said.

He answered, "Shucks Mister. What's time to a hog?"

And so it goes for BBQ'ers who choose to cook a whole hog. Whether for competition or just back yard braggin' rights; it's hard to beat the overall impression and the downright awe-inspiring effect of presenting a well cooked whole hog to family and friends or competition judges. It takes an unusual amount of patience and stamina to balance the demands of 24 hours of fire management, basting, aspiration and alcohol consumption that normally accompanies smoking a whole hog. But, when done well, few other outdoor cooking experiences come close.

Hog selection is limited more by access to a local slaughter house or offerings of local distributors. Fresh is always better than frozen and smaller (younger) is better than larger (older). You may have to ask your local butcher to special order your hog. As I mentioned earlier, Holy Smokers Too used to be able to go into the coolers at Fineberg Packing and pick our hogs. You probably will not have that opportunity due to current restrictions by the USDA. Memphis in May (MBN) rules specify a hog of at least 85 pounds.

Holy Smokers Too generally chooses a 120 pound hog that has been freshly killed. Before Louis Fineberg died, and the USDA cracked down on rules on access, we used to select our hog from the Fineberg slaughter house inventory. Louis always made sure we got a female with one blue eye and one brown eye and we named her Oink Johnson. In most cases, purchasers have to rely on the staff at the packing house. On rare cases, a competitor shows up with a farm-raised hog of his own, but not too many contests allow non-USDA inspected meat products. What you do in your own back yard is up to you.

Ask your processor to split the back bone for you. They can do this in the plant with a power saw that is quicker and more sanitary than the sledge hammer and axe we used to use. You might consider having the ribs separated from the backbone by slicing down both sides of the backbone as part of this process. This will allow the hog to lay flat as well as expose some of the loin to smoke and make the ribs easier to remove when the cooking process is completed.

The hog should be trimmed to further remove any excess fat or blood clots. Excess skin should also be removed. This also includes the "skin" on the rib cage. This is the same membrane we want to remove when cooking the ribs individually. This is one of the serous membranes that hold and protect internal organs in the hog. The membrane on the ribs is the pleura membrane – for you scientific types. Some competitors will remove all of the hog's external skin. Remember, the more muscle surface area exposed – the more flavorful "barq" that will be produced and the greater smoke penetration is possible. You have to weigh the risk of lost moisture vs. smoke penetration and "barq" formation. I know of at least one team that removes the entire skin as intact as possible so that they can then replace the skin during the cooking process to protect from moisture loss. That team also coats their seasoned hog with brown sugar prior to replacing the skin. This creates a caramelized coating on the whole hog. Like I said, "Some like it sweet!"

We generally trimmed the already exposed sections of the ham and shoulder areas of excess fat. Then, a knife or fingers were used to separate the skin from the muscle as far down as the shank in both ham and shoulder. This allows us to rub our dry rub deep under the skin in those areas. The entire carcass skin should be wiped down with a damp cloth as a final cleaning process in preparation for cooking.

Spray the exposed meat and ribs with a coat of apple cider to add a little sweetness and help the dry rub adhere. Dry rub should then be applied generously to all exposed parts of the hog. A coating of vegetable oil rub on the skin will also reduce drying and cracking as well as enhance the skin's color while cooking.

One of the most successful hog teams likes to inject their hogs with fruit juices. They will inject their hogs before the cooking process, at a mid-point, and then at the end of the cooking process when the muscle tissue is breaking down so they get a good distribution of juice.

You can also use the hog's circulatory system for injecting by inserting your syringe into the brachial and femoral artery in the shoulders and hams. This takes approximately 40 psi of pressure to be affective, so don't be surprised if the majority of your liquids squirt back out of the artery.

During our *Decade of Dominance,* we always started the hog belly-side-down for the first six hours. This maximized smoke exposure, rendered excess fat and sealed the exposed muscle with the beginning of the new skin or "barq". After the initial phase, the hog was flipped and the process of hourly basting began.

Your smoker should be brought to 300 - 350° before placing the hog in the smoker. Once the hog is inside, the temperature should be allowed to drop down to the 225 - 250° range for the next 24 hours. Your target for hams and shoulders should be an internal temperature of 185 - 190° in order to allow you to pull your Que from the hog.

You should maximize your smoke for the first 6 – 8 hours. After that, the amount of smoke flavor the hog will accept diminishes. *See the section on the smoke ring.* Since you are looking at an extended cooking period, a moist cooking chamber may help keep your hog from drying out. This is when a moisture pot of juices, apples, and/or secret elixir may make sense. We use a cast iron pot place against the coals to keep a gentle steam in the cooking chamber. Whatever combination floats your boat. The key is to keep just enough moisture in the air to reduce the drying affect of the smoking process.

Once the desired internal temperature is reached, you can allow the smoker to drop to around 170° to keep the hog warm until your guests arrive or judging begins. There will also be an accumulation of baste in the hog's chest cavity at the end of the cooking process. This should be aspirated or remove as best you can in the last hour of the cooking process. This will allow the surface areas to dry out and allow for a final dusting of dry rub or glazing with sauce for appearance and added flavor.

Some teams use a blanket made of aluminum foil to cover their hogs during the period after smoking has ceased in order to preserve moisture during the extended cooking time. If it works for shoulders and ribs; why not your whole hog?

We have found the hog weight of 120 pounds and average cooking temperature of 225° works well for our team's temperament and a 24 hour cooking schedule. You'll have to experiment to see which works best for you. Hogs of 350 pounds and cooking times of 9 hours have been witnessed at contests. Although, rarely by the same team.

During judging, it's best to pull your preliminary blind entry and on-site samples from the side of the hog that is further from the opening of your smoker. This preserves the front of the hog for your presentation to the finals judges. *You were planning on making it to the finals, right?*

Billy Bob's Basic Rub

2 tablespoons	Sweet Basil	1 tablespoon	Garlic Powder
2 tablespoons	Coarse Black Pepper	1 ½ teaspoons	Celery Seed
2 tablespoons	Paprika	1 teaspoon	Dry Mustard
1 tablespoon	Chile Powder	2 teaspoon	Ground Cumin
½ tablespoon	Red Pepper, Cayenne		

Dust all areas of exposed meat and areas beneath the skin around shoulders and hams.

Whole Hog Baste

| 1 gallon | Apple Cider Vinegar | 1.25 cup | Lemon Juice |
| 1 quart | Worcestershire Sauce | 3 tablespoons | Black Pepper |

Combine and bring to a hard boil 10-15 minutes. Baste hog on the hour until the last hour when you should aspirate the body cavity and remove the liquids that have pooled over the ribs. Allow cavity to dry out during the last hour.

Barbe-a-que – "Whiskers to Tail"

On the way to Kennett, MO for the Missouri State Championship one year, I noticed a hog farm with a pen close to the highway. In a pen, separated from the other hogs, was a large hog that must have weighed over 450 pounds. I thought that here was a hog that would yield extremely large shoulders, ribs and other parts, but would probably be too tough for a competition entry. What grabbed my attention was that the right front leg appeared to have a homemade pegged leg attached to the hog's shoulder. I made a mental note to check him out on my next passing. Two days later, with no prizes or awards from the Kennett contest, I decided to stop by and check out this pegged leg pig.

I pulled into the farmyard, got out of my car, and approached the farmer who was working on a broken-down tractor.

I said, "Tell me about this hog with the pegged leg."
"You mean George the Wonder Hog?" he asked.
"George the Wonder Hog," I queried?

He went on to tell me that his grandson had been visiting last summer and fell into the well. This big hog heard the boy yelling for help and broke out of his pen. He then pushed the water bucket into the well and pulled on the rope until he raised the boy from the bottom of the well. He saved the boy's life.

I said, "That's a great story, but tell me about the pegged leg."
The farmer said, "Hell, mister. I hog that special can't be eaten all at one time!"

The following pages highlight champions who have proven time and time again that some hogs are too special to be eaten all at one time and have shown that sum of the parts is often greater than the whole.

Pork Shoulders and Boston Butts

In the world of pulled pork, no cut of meat is as forgiving as a pork shoulder. The full shoulder containing the Boston butt, picnic shoulder and shank can run from 10 -20 pounds or larger. It's a complex muscular cut with ample fat. Cooked low-and-slow a shoulder will yield a moist, tender flavorful pan of smoked pork.

Ernie Mellor, Pit Master of the Hog Wild competitive BBQ team gave me the following testimony to the versatile shoulder. In the 2001 MIM World Championship contest, Ernie left his team in the middle of the cooking process of his shoulders to go to his room and freshen up. When he returned, he discovered there had been a grease fire in the smoker giving his shoulders the appearance of large bowling balls. As he cut through the glossy black surface of the shoulders to assess the damage, he discovered that the meat in the center of the shoulders was unharmed and actually tasted quite good. Being the experienced Pit Master and pitchman that he is, Ernie came up with a plan. When the judges came to the Hog Wild tent for judging, this is part of what he said, "I don't know how we did it, but we were able to produce the darkest mahogany skin on our shoulders than ever before. We think we have a winner." He then pulled the center-cut pork for tasting and the end result was a second place finish in the shoulder category for 2001. Not bad for the now infamous Hog Wild bowling ball.

We have previously talked about bringing the shoulder to 180° or more in order to tenderize the muscle fibers so your pork can be "pulled" for serving. I mentioned that North Carolinians like to pull their butts from the cooker around 160° and chop it into a fine hash. *I know, and truly hope, that the North Carolinians reading this will disagree and say they smoke their pork to the pulling stage. I'm simply stating what I have observed at contests in N.C.*

Shoulders should be trimmed of excess fat and rinsed of blood clots. For a whole shoulder, the skin cap and fat on the shank are normally left in place. In preparation for cooking your shoulders or butts, you can apply rubs, liquid marinade or inject the shoulders for up to 24 hours in advance. Apply the rubs and wrap your shoulders in foil or plastic wrap and place in the refrigerator (or ice chest) for the marinating period

If you are going to inject your shoulders or butts, I suggest you inject prior to placing your meat in the smoker and then again prior to wrapping your meat. If you are not wrapping, then a second injection when your meat passes the 170° mark will probably work best.

Shoulders will cook from 10 to 16 hours or up to 24 for large cuts, therefore it's not necessary to bring them to room temperature before placing on the smoker. I generally like to dust my shoulders one more time before placing on the smoker. Shoulders dusted the night before will have formed a paste on the surface from the dry rubs mixing with proteins and juices brought to the surface.

I place shoulders and butts fat-side-down at the beginning of the cooking process. I tell my students to do the same for two reasons: (1) I'm not always sure how well I've (they) cleaned the cooking surface and I like the fat to serve as a protective coating as it melts off the meat, and (2) during the initial basting period (6 hours) I want to be developing layers of flavor with the baste and *barq* that is forming, without having to turn the shoulder over for basting and back again, scraping the surface against the smoker's surface grills.

When the shoulders have cooked for six hours and reached a temperature of at least 160°, I like to wrap them in aluminum foil for the final process. Once the shoulders have reached 180° they have reached the pulling stage. Depending on the size and age of the shoulders it may be necessary to keep them at 180° plus for an hour or more to reach your desired tenderness.

Both pork shoulders and Boston butts will tell you when they are ready. Like the pop-up indicators in your Thanksgiving turkey your pork has a natural gauge. For the full shoulder, simple grasp the shank bone and twist. If your pork is ready it should release easily and pull from the shoulder. The same for the blade bone in the Boston butt. Both bones should come out relatively clean and begin to marbleize as an indication the meat is cooked to the proper stage.

In the final wrapped stage, you may choose to turn the meat fat-side-up so that the meat self-bastes during the wrapped process. If so, it is a good idea to remove the meat from the foil and place it on the smoker fat side down for the last 30 minutes to dry out the surface (if it's not so tender that it falls apart). Before you remove the foil, poke a hole in the bottom of the wrapped foil and allow the juices to drain out. This should be done carefully since the juices will be very hot. Some competitors will then glaze the shoulder with sauce or apply a final dusting with dry rub prior to serving. It's your choice.

HOG WILD – SHOULDERS

Estimated cooking time is 16 – 18 hours for shoulders (15 – 16 lbs) and 10 – 12 hours for Boston butts (6 – 8 lbs).

Ernie Mellor, Pitmaster of the HOG WILD competitive BBQ team and catering company in Memphis, TN likes to inject and marinate his butts and shoulders with his own HOG WILD *Hog Wash* marinade. Depending on the size of the cut and time available he likes to let them marinate from 6 – 24 hours.

He recommends cooking at 225 degrees. He also rubs his butts with HOG WILD *Special Dry Rub* prior to placing on the smoker. He recommends basting when you're adding fuel.

In about the 8th hour, carefully place the shoulders/butts in aluminum foil boats. Add about 2 inches of liquid to the boat (water, beer, marinade, or *Hog Wash*). After 3 to 4 hours of cooking in the boats, let the liquids out of the boat.

Internal temperature must reach 180 degrees for two hours to meet his tenderness requirements. Ernie says to pull the meat in pieces the size of your index finger being careful to remove any remaining fat, cartilage or gristle.

Before serving he likes to lightly sprinkle the pulled pork with his *Special Dry Rub* and toss it like a salad.

I agree with Ernie that the meat should stand on its own, so he serves his with sauce on the side.

HOG WILD cooking tips, Hog Wash and other products can be found at www.hogwildbbq.com.

HOG WILD has won many a trophy and supplied first class Que to many an event around the mid-south, including the "Bowling ball" that won him a second place finish in the shoulder category at the 2001 Memphis in May World Championship BBQ Cooking Contest.

Ribs, Ribs, Ribs

Billy Bob Billy's
Bombastic BBQ Ribs

Baste

108 oz. BBQ Sauce
108 oz. Apple Cider/Apple Juice
28 oz. Wickers Baste
28 oz. Lemon Juice

Basic Rub		Finishing Rub	
2 tablespoons	Sweet Basil	2 tablespoons	Sweet Basil
2 tablespoons	Coarse Black Pepper	2 tablespoons	Coarse Black Pepper
2 tablespoons	Paprika	2 tablespoons	Paprika
1 tablespoon	Chile Powder	1 tablespoon	Chile Powder
½ tablespoon	Red Pepper – Cayenne	½ tablespoon	Red Pepper
1 tablespoon	Garlic Powder	1 tablespoon	Garlic Powder
1 ½ teaspoons	Celery Seed	1 ½ teaspoons	Celery Seed
1 teaspoon	Dry Mustard	1 teaspoon	Dry Mustard
2 teaspoons	Ground Cumin	2 teaspoons	Ground Cumin
		4 tablespoons	Brown Sugar

Baby Back (loin) ribs 2.25 lbs and down. Allow ribs to reach room temperature. Remove the pleura membrane from backside and trim excess fat and rough cuts for appearance. Rinse any blood spots and pat dry. Spray ribs with apple cider and apply basic rub. If you are going to allow ribs to marinate overnight, you need to lighten application of rub accordingly; otherwise I prefer a solid coating of rub followed by another cider spraying. Allow rub to sit for one hour before placing ribs on cooker.

Preheat smoker to 300° - 350° (temp. will drop considerably while you are placing the ribs in the cooker). Place ribs bone-side-down with enough spacing to allow air circulation. Apply wet wood chips to your charcoal to produce heavy smoke. Smoke ribs for one hour at 220° - 250°. After one hour, baste ribs front and back liberally with baste. Smoke for thirty minutes and repeat basting. Smoke for thirty more minutes and repeat basting. After two hours, wrap ribs bone side down in loose tent-style aluminum foil with approximately ¼ cup of baste or apple cider inside foil. Cook for one more hour with ribs wrapped in foil. At the end of the third hour, remove ribs from the foil and place on the smoker. Spray with apple cider and smoke for additional 30 minutes to dry ribs. After 30 minutes, dust ribs with finishing rub and continue cooking for 30 minutes for Memphis-style dry ribs. *If you prefer wet ribs, brush on BBQ sauce instead of finishing rub at this point and then again just before removing from smoker. If ribs are heavier and appear stiff when you are prepared to remove from foil, allow them to tenderize for another thirty minutes before removing from the foil for the drying stage.*

Apple City Championship Ribs

Dry Rub

10 tablespoons	black pepper		5 tablespoons	cayenne pepper
10 tablespoons	paprika		2 tablespoons	celery salt
5 tablespoons	garlic powder		1 tablespoon	dry mustard
5 tablespoons	chili powder			

Baste
Apple juice (warmed)

Finish Sauce

32 ounces	Hunt's Ketchup		1 tablespoon	garlic powder
8 ounces	apple cider vinegar		1 tablespoon	white pepper
8 ounces	soy sauce		1 medium	onion (finely grated)
4 ounces	Worcestershire		1 grated apple	Golden Delicious
4 ounces	apple juice		¼ grated	bell pepper

Allow ribs to reach room temperature. Skin ribs and trim for appearance. Rub both sides with dry rub. Wrap ribs and place in refrigerator for 4 to 10 hours. Bring first 7 sauce ingredients to a boil. Stir in grated ingredients and simmer until sauce will coat a spoon.

Smoke ribs for 5 to 7 hours in a 180 to 200 degree range. Baste hourly with warm apple juice. Use soaked apple wood chips for smoke. During the last 30 minutes, glaze ribs with the finish sauce. Sprinkle with dry rub before serving with sauce on the side.

As I mentioned before, we went head-to-head with the Apple City Team as the first two teams to have the chance to be the first three-time World Champions at the Memphis in May BBQ Fest in 1994. We won in Whole Hog and they won in Ribs. The rest is history. The Apple City gang went their separate ways after 1994 and the Holy Smokers Too became more of a "social" team. *We never did take losing very well.*

BIG BOB GIBSON
BAR-B-Q RIBS

2 SLAB RIBS
¾ Cup Big Bob Gibson Seasoning and Dry Rub

Marinade
1 cup apple juice
1 cup grape juice

Finishing Glaze
¾ cup Big Bob Gibson Championship Red Sauce
¼ cup Honey

Pull membrane from back of the ribs. Sprinkle ¼ cup of rub over each slab, front and back. Place ribs, meat side up, on the grill at 250 degrees for 2 hours. Remove and place each slab on a double aluminum foil square, meat side down. Pour 1 cup of marinade mix over each slab while wrapping lightly in foil. Cook for 1 hour at 250 degrees. Remove slabs from foil and sprinkle with remaining ¼ cup of rub. Cook an additional hour uncovered, until ribs are tender. Remove, paint with finishing glaze, and cook for 15 minutes. Cut and serve.

www.bigbobgibson.com

I can't speak for Chris Lilly and his team's ribs, but they did win the shoulder category an unprecedented 6 times in a row at the Memphis in May World Championship BBQ Cooking Contest – so you can be sure the boy knows his Que. Big Bob Gibson's BBQ restaurant has been in operations since 1925. Contests were just another way of promoting their restaurants and catering business until Chris married into the family. They got a real showman and Chris got access to great BBQ. His wife Amy (Big Bob's great-granddaughter) is also easy on the eyes. The two go together like BBQ chicken and white sauce.

Hogaholics Mouth-Watering Ribs

Hogaholics Dry Rub

2 tablespoons	coarse salt	1 tablespoon	chili powder
2 tablespoons	sugar	1 tablespoon	paprika
1 tablespoon	lemon zest	½ tablespoon	black pepper
1 tablespoon	garlic powder	½ tablespoons	cayenne pepper

Hogaholic's Baste

4 cups	Wickers Marinade and Baste	2 cups	vegetable oil
2 cups	apple cider vinegar	1/2 cup	lemon juice

Allow the ribs to reach room temperature. Skin the ribs and trim for appearance. Rub both sides with dry rub. Place the ribs bone-side down on the smoker, away from coals. Smoke the ribs for 1 ½ to 2 hours at 225°. Baste every hour for the next 3 ½ to 4 hours. If you must, glaze with BBQ sauce in the last 30 minutes.

The Hogaholics were always one of my favorite rib teams. Team members with a FedEx connection would fly in fresh salmon that we would cook on our larger smoker (keeping one for our trouble). Although they were rarely on the stage at Memphis in May, they always served great mouthwatering ribs and a king's portion of hospitality. They were very successful in other contests including Grand Champions in 1985, 1986, and 1987 at the Knoxville River Feast, but (as many teams have) they finally retired from competition. Team member Roger Sapp continues to compete with a new team.

Smokin' Friars Espresso Ribs

Espresso Rub

¼ cup	espresso beans (finely ground)	2 teaspoons	kosher salt
¼ cup	Ancho chili powder	1 tablespoon	ground black pepper
2 tablespoons	Spanish paprika	1 tablespoon	ground cumin
2 tablespoons	dark brown sugar	1 tablespoon	dried oregano
1 tablespoon	dry mustard	2 teaspoons	ground ginger
		2 teaspoons	cayenne

Mustard-Vinegar Baste

3 tablespoons	olive oil	2 tablespoons	sugar
1 medium	Red onion	2 teaspoons	mustard seeds
2 cloves	garlic, finely chopped	2 teaspoons	coriander seeds
1 ½ cups	cider vinegar	¼ cup	Dijon mustard
½ cup	water	1 tablespoon	Worcestershire sauce

Heat the oil in a medium saucepan over medium heat. Finely chop onion and garlic and add to saucepan. Cook, stirring occasionally, until soft – about 3-4 minutes. Add the vinegar, water, sugar, mustard and coriander seeds and bring to a simmer. Cook for 5 minutes, then remove from the heat. Wisk in the Dijon, dry mustard and Worcestershire. Let cool to room temperature. The sauce can be made up to one day in advance.

Ribs

2 racks pork ribs, about 2 ½ lbs. each
2 tablespoons canola oil
¼ cup thinly sliced green onions, white and green sections

Brush the ribs on both sides with the oil and apply the rub. Smoke for 2 ½ hours, Begin basting, continue to baste every 15-30 minutes until done. When the ribs are tender, remove from smoker, apply more baste, cut as desired and garnish with green onions.

This recipe came from Eric Harralson of the Smokin' Friars, Knoxville, TN They are active with the Episcopal Church and host many fund raisers during the year. On occasion, they will cater other local parties in the Knoxville area. Eric is one of my successful students from the University of Tennessee BBQ classes.

Pork Loin

Pork loin is one of the most versatile and economic cuts of pork available. Versatile, because its mild flavor readily accepts spices and economic, because there is very little waste. But, because it is so lean, it can be overcooked to a tough stage if you are not careful.

Here is a simple recipe we used on an outside fireplace in Perugia, Italy with herbs and spices we found around our vacation villa. www.celestinabbey.com. I have slightly modified the process for your smoker.

One 4 pound pork loin

½ cup	chopped lavender
½ cup	chopped rosemary
4 cloves	garlic
2 tablespoons	mixed peppercorns, cracked
1 tablespoon	salt
EVOO – a good quality extra virgin olive oil	

Roughly chop the garlic cloves and then massage into a paste using the salt as an added abrasive. Finely chop several handfuls of the lavender leaves and rosemary until you have approx. ½ cup of each. Sprinkle the olive oil over the pork loin and smooth the coating with your hands as you add the garlic paste. Sprinkle the lavender, rosemary, and cracked pepper on your cutting board and roll the loin until evenly coated.

We cooked the loin on the outside area of the fireplace approx. one hour, turning ¼ every 15 minutes to brown the skin, until medium rare.

At home, I would place the loin on the smoker at 225°and smoker for approximately 2 hours or until an internal temperature of 145°. I would baste on the half hour. The Smokin' Friars' Mustard Vinegar Baste would be a good choice with this dry rub.

Any baste will add unique flavor to the pork loin, so I suggest you try your own combinations of marinades, dry rubs, and bastes until you find a favorite. With the versatility of the pork loin, there is no reason you should get bored. Just, be sure and avoid overcooking this cut. Leftovers, the next day, make a great sandwich when sliced in deli-thin servings. Remember, you can always put it back on the smoker if it needs more, but it's hard to save a piece of meat that's overcooked.

Beef Brisket Recipes

Billy Bob Billy's Beef Brisket

Billy Bob Billy's
Best Brisket Rub

This rub takes a time to prepare, but the fresh-ground spices and the extra effort are well worth the results.

2	dried chipotle peppers	¼ cup	sweet paprika
2 tablespoons	black peppercorns	1 tablespoon	garlic powder
1 tablespoon	cumin seeds	1 tablespoon	sweet basil
1 tablespoon	coriander seeds	1 tablespoon	onion powder
1 tablespoon	mustard seeds		

Toast seeds in a dry cast iron skillet over medium heat (2- 4 min). Grind seeds, peppers and peppercorns with mortar and pestle into coarse powder. *Those of you who are privileged can use your electric spice grinders, but don't overdo it. You want a coarse grind with explosive bits of flavor, not a fine powder.* Combine all ingredients and apply liberally to your brisket.

Billy Bob Billy's
Basic Baste & Marinade

1 quart	Kraft Original BBQ Sauce	1 cup	Wickers marinade
1 quart	apple cider	1 cup	lemon juice

Select a 6 – 8 pound brisket or ½ brisket cut. Trim the fat down to ¼ to ½ inch and remove connective membrane, if visible. Place brisket in plastic bag with ample marinade and leave in refrigerator for 4 – 6 hours or overnight. Remove brisket from marinade, pat dry and add dry rub. *Apply optional mustard pack if you do not want to baste.* Allow dry rub to rest on brisket for one hour. Preheat smoker to 275° to 300°. Place brisket in smoker fat side down. Maintain a cooking temperature of 225°, basting the brisket every hour. After 6 hours and an internal temperature of at least 160°, wrap brisket (fat side up) in loose aluminum foil tent with a splash of baste. Continue cooking at 225° for four hours or until internal temperature of brisket has been 180° or higher for one hour. Remove brisket from foil and place in smoker (fat side down) for 30 minutes to dry surface area. Remove from smoker and allow brisket to cool to 150° degrees for slicing in ½ inch thick slices for serving. Leftovers can be sliced deli-thin while cold for great sandwiches served with Dijon mustard.

Billy Bob Billy's
BBQ Brisket Chili

4 – 5 pounds	smoked brisket	1 tablespoon	masa harina flour
6 slices	center cut bacon	1 teaspoon	oregano
6	poblano peppers	1.5 teaspoon	kosher salt
2 cups	roasted tomatoes	2	bay leaves
1 medium	onion, diced	1 small can	green chilies
3 cloves	garlic, minced	1 bottle	dark beer (Guinness)
2 tablespoons	Chile powder	1 fresh	habanero pepper

This recipe is great to rescue that tough cut of smoked brisket that did not turn out like you planned. *Although it's much better with brisket smoked just for this chili.* Refer to the section on smoking brisket in order to prepare your brisket. You want to carry the brisket to the wrapping stage (approx. 6 – 8 hours of smoking). At that point, carve out the 4 to 5 pounds you will use for chili. Reserve the brisket to be cut into one inch cubes to be added to the chili later. If you aren't smoking brisket, you can brown fresh brisket for this recipe.

Prepare a medium-high fire in your grill for direct grilling of the peppers and tomatoes. Place the 6 poblano peppers and 6 – 8 plum tomatoes on the grill. Turn the peppers and tomatoes until the skins on the peppers are blackened. Remove the peppers and put them in a paper sack or plastic bag to steam the skins. Remove the tomatoes and set aside to cool. After about 15 minutes, remove the peppers from the bag and peel off the skins.

Remove the seeds and cut the peppers into strips and then cut the strips in half. Allow the tomatoes to cool and then, remove the skins and dice the tomatoes. If you are a little pressed for time you can cut the tomatoes in half and squeeze out the meat and discard the skins. Or you can place them in a blender and puree skins and all. I prefer a little more tomato texture, so I prefer to cube the peeled tomatoes. Cutting the tomatoes in half before placing them on the grill will impart more smoked flavor to the meat if you plan to remove the skins.

In a large pot, crisp up the bacon after slicing into small pieces. As the bacon bits begin to brown, add in the diced onion and cook until translucent. *If you have raw brisket, remove the bacon bits and onions and reserve until later. Sauté the cubed brisket in the bacon drippings until browned on all sides. Then return the bacon and onions.* Remove the seeds and membranes from the habanero pepper and cut it into micro-julienne slices. *Suzie Que and I call these carrot slices.* If you are not into this much heat, you can substitute two jalapeno peppers.

Once the onions are translucent, add the smoked brisket cubes and all of the other ingredients except the beer and masa harina into the pot. Bring the chili to a boil and then lower to simmer for the next two hours. After one hour add the bottle of beer and stir in the masa harina as you sprinkle the flour into the chili so that it does not clump. The masa harina will thicken the chili – if you want a thicker chili, you can double the masa harina. Simmer the chili for a minimum of one more hour, testing the brisket cubes for tenderness.

Serve in a large bowl with a dollop of sour cream or a sprinkle of Mexican three cheese shredded cheeses. Suzie Que likes it served over a bed of Fritos. A Memphis favorite was to thicken the chili and serve it over Fritos with a scoop of Cole slaw in the middle. The Cole slaw balances the heat of the habanero peppers nicely. It also goes well served over a scoop of steamed rice. Pick one or develop your own style.

Make the chili a day in advance to allow the flavors to meld. *Or enjoy while your BBQ brisket continues to smoke.*

Serves 4 – 6 people.

HOG WILD – Brisket

Ernie Mellor of the Hog Wild BBQ team offers this process for brisket.

Start with a 6 – 8 pound brisket.(*Probably a half-brisket*) Ernie says, "If there is more than 2 to 3 inches of fat on the brisket, trim it down. You must marinade the brisket at least 12 hours, and then rub, rub, rub that puppy down with a good dry sauce."

I love a man that becomes one with his meat. My son-in-law says, "It's not the size of your smoker, it's how you rub your meat!"

The brisket should then sit for an additional 2 – 4 hours. Place the meat fat side down in the cooking' chamber with a temperature of 250 – 275 degrees the first hour to hour and a half. Then maintain 225 degrees.

"You need only baste when you're messin' with your fire," says the Pitmaster.

In the 7th or 8th hour, baste it heavily and wrap tightly in foil. Cook an additional 2 – 3 hours. The brisket is done when the internal temperature reaches 180 degrees.

IMPORTANT: The brisket must cool down to 150 degrees plus/minus before you can slice it. Always cut a brisket against the grain of the meat.

Anything Butt Recipes

Billy Bob Billy's
Barbecued Bologna

Pick up a 10-pound stick of bologna.

Peel the casing from the bologna and slice the stick in 2-inch rounds. The rounds will optimize the surface area and allow you to build more barq and more smoked flavor than cooking it as a 10 pound stick.

Dunk the rounds in your favorite baste and place in the smoker. You can either grill for 30 minutes, turning at 15 minutes, or smoke the bologna until it has developed a dark "bark" on both sides (1 hour).

Cut the rounds into one inch cubes and dust with your favorite dry rub. Serve as an appetizer with a mustard dip or Dijon. Don't be concerned if some misguided guest tells you they like the bologna better than your Que.

If so, it's time to revise your guest list – or go with the flow and pour another cold one.

Mustard Dipping Sauce

1/3 cup	brown sugar	1/4 cup	mustard
1/4 cup	onion, finely chopped	1/2 teaspoon	celery seed
1/4 cup	vinegar	1/4 teaspoon	garlic powder

For the sauce, in a saucepan combine brown sugar, onion, vinegar, mustard, celery seed and garlic powder. Bring to a boil, stirring 'till sugar dissolves. Allow to cool and place on table as a dipping sauce for your bologna.

HOG WILD - Barbequed Bologna

"Get your hands on a stick of bologna, 3 – 7 pounds," says Ernie.

Take your knife and make a cut the length of the loaf. This cut should be about half way through the loaf. Now, peel the casing off the bologna. Roll the loaf in any marinade, something to get it wet. Be sure the marinade gets in the crack you just cut. Sprinkle that "thang" down with your rub. Place the loaf in the coolest section of the cooker (200 degrees +). As the bologna cooks, the slit you made will open up wide. This should take 2 to 3 hours. Slice the bologna every ½ inch and grill them over direct coals or the hottest part of you smoker for another 10 minutes.

"Paint 'em with warm barbeque sauce, and slap 'em on a bun with cheese and onion. Look out, this thing is real," says Pitmaster, Ernie Mellor of the Hog Wild team.

Billy Bob Billy's
BBQ Pizza

1	pre-made 12" pizza crust	½ cup	chopped green onions
1 cup	BBQ sauce	¼ cup	chopped fresh cilantro
1 cup	pulled pork BBQ	3 sliced	jalapeno peppers
2 cups	3 cheese Mexican shredded cheese		

1	18" pizza stone

Preheat grill and pizza stone with hot charcoal to approx. 450° (medium high).
(If you don't have a pizza stone, a heavy pan covered with aluminum foil can be used. If not, place a double thickness of aluminum foil on the grill at the time you insert your pizza).

Cover the pizza dough with BBQ sauce. Add the pulled pork and cover with cheese and green onions. Place jalapenos strategically around the pie. Transfer to the cooker and cook until the cheese melts and the crust is golden – 8 to 10 minutes. Slice and top with chopped cilantro.

To kick it up a notch*. Micro-julienne a cored habanero pepper and sprinkle at your own risk. For an extra-crispy pizza, pre-cook the pizza for five minutes to stiffen and brown the dough before adding the toppings for the final five minutes.*

Cilantro and Lime Cedar Plank Salmon

2 Salmon fillets 8 – 10 ounces each

Salt and pepper
2 limes
Olive oil
1 cup cilantro leaves

2 Cedar planks (1 x 4 untreated cedar in 8 – 12 inch lengths)

Soak cedar planks in water one hour plus.
Preheat smoker to 325 degrees.

Mix juice of two limes and olive oil in a mixing bowl to form a smooth marinade. An approximate mix of 3 parts oil to 1 part lime juice should do. Mix in ½ of the cilantro leaves. Salt and pepper to taste. Coat the fillets with the marinade mixture, place in plastic bags or plastic wrap, and refrigerate for 1 hour.

Remove planks from water and allow them to drain. Spray or brush planks liberally with olive oil on smooth side of plank. Place planks in grill to preheat for approx. 10 minutes. Remove planks from smoker and build a bed of cilantro on each plank. Place each fillet on a plank and drizzle with any marinade left in the plastic. Place in the smoker away from heat source. Smoke with mild smoke (apple, pecan, etc.) for 12 to 18 minutes until fish flakes with a fork.

Serve on planks for affect or transfer to serving dish making sure to transfer as much of cilantro bed as possible. To reduce clean up, the cedar planks may be placed on a cookie sheet prior to smoking to catch any juices that may seep from the salmon.

NOTES

Birds of a Feather Recipes

Driving thru Mississippi for a contest many years ago turned me off of smoking chickens for quite a while. I was driving down this back road when a chicken running down the road passed me like I was sitting still. And I was doing 65 miles per hour! I followed the chicken until it turned down a driveway and ran into the barn behind an old farmhouse. I pulled into the drive, parked my car, and walked up to the farmer whittling on the front porch.

"Tell me about that chicken that ran into the yard. Why is it so fast?" I asked.

"Well," said the farmer, "Me and Maw and Junior was always fighting over the drumsticks when we had our Sunday chicken dinner."

I said I understood, because I was partial to legs myself. He told me that Junior had gone off to school at Mississippi State University studying biology and animal husbandry. When he came home he brought with him a three legged chicken he had bred by splicing the cells in a chicken embryo.

"We was real excited that we could all have a drumstick thanks to Junior's work," said the farmer.

"That's fascinating," I said. "How does it taste?"

"Don't know," said the farmer. "We can't catch him!"

Yard birds, either chickens or game hens, are best barbecued on the grill, in my opinion. There is a tendency for chicken to dry out over long periods of cooking and a tendency towards rubbery skin at low temperatures and the use of vinegar based marinades. There are a couple of chicken grill recipes in the *Great Grill Section.* If you are determined to cook chicken on your smoker, you must crank the heat up above 350° over the final thirty minutes to an hour to crisp up the skin or serve your bird naked.

Beer can chickens were first introduced at Memphis in May by a cooking team from the Air Force Reserve in Memphis. Steve Raichlen has written an entire book on beer can chicken. I see no need to add another redundant chapter on beer can chickens. Pick up Steve's book or any other BBQ book and you will find ample examples. I have included one duck smokin' recipe and a favorite chicken recipe of mine that I like, but judges have not.

Billy Bob Billy's
Make Your Mark
Chicken Thighs

8 – 10 Chicken thighs

Marinade		**Dry Rub**	
1 quart	BBQ Sauce	2 tablespoons	Sweet Basil
1 quart	Apple cider	2 tablespoons	Coarse Black Pepper
1 cup	Wickers marinade	2 tablespoons	Paprika
1 cup	lemon juice	1 tablespoons	Chile Powder
½ cup	Makers Mark Bourbon	½ tablespoon	Red Pepper / Cayenne
		1 tablespoons	Garlic Powder
		1 ½ teaspoons	Celery Seed
Glazing Sauce		1 teaspoon	Dry Mustard
		2 teaspoon	Ground Cumin

2 cups BBQ Sauce
1 shot Makers Mark Bourbon

Dust your thighs with dry rub, including under the skins. Place thighs in plastic bags with marinade for one hour or overnight in cooler or refrigerator. One hour before cooking, pull thighs from refrigerator or cooler. Debone ribs, taking care not to cut the skins. Reform the thighs into oblong balls using the skin as a wrapper. Dust with dry rub and place in the smoker at 225°. Cook for 1½ hours without touching the thighs. At the 1 ½ hour mark, baste with the glazing sauce for two 15 minute intervals. At two hours (or 140° internal temperature) transfer the thighs to a medium hot grill. Glaze the thighs while turning often in order to crisp up the skin and caramelize the glazing sauce.

The alcohol will burn off the sauce,
so you may need to pour yourself two fingers of bourbon before tasting!

Bob Billy's
BBQ Cornish Hens

Another marketing plan revealed! Cornish hens do not come from Cornwall nor are they "wild game" or "game hens". It's just a way to add some exotic twist to the cooking of young chickens. Using the term "baby chickens" would be a real sales downer. But, let's move on. The process for cooking small hens can be the same used for chicken fryers, it just takes more time the larger the bird.

You can use your favorite dry rub and baste/marinade. I have had great success using my basic dry rub and baste.

Thoroughly clean the hens and remove any organs packed in the birds. After washing pat them dry before applying the dry rub. Apply your initial dry rub both on the inside and the outside of the hen. Allow the hen to rest for 30 minutes to an hour. Place each hen in a gallon freezer bag with one to two cups of the marinade you have chosen. Store the hens in the refrigerator for two hours or overnight. Visit the birds several times during the night (or day) and turn them over several times to make sure they are being coated by the marinade. One hour before placing on the smoker, remove them from refrigerator and allow them to come to room temperature.

Bring your smoker to 350° prior to cooking. Give your birds a final dusting with dry rub and place on the smoker backside down. Apply wood chips for smoke and cook at 275° to 300° until you get a reading of 160° from your thermometer placed inside the thigh or breast. The legs should be very loose as the natural gauge of doneness. Baste the birds every 30 minutes to develop layers of flavor on the skin. This should take approximately 2.5 – 3 hours.

Full sized chickens or fryers will take 4 hours to reach the proper temperature. For larger birds you can push dry rub, herbs and butter under the skins to add flavor.

If you get to the done stage and the bird's skin is rubbery and not crisp, you should crank the heat up to 400° or transfer the bird to a medium high grill to crisp up the skin, turning the bird frequently over 30 minutes or more.

Bob Billy's
Smoked Turkeys

One 10 -12 pound turkey

Turkey Brine

1 cup	coarse salt	4 twigs		thyme
1 cup	sugar or honey	6		bay leaves
6 twigs	rosemary	2 gallons	water	

Dry Rub

4 tablespoons sea salt
4 tablespoons black pepper
4 tablespoons sweet basil
4 tablespoons thyme

4 tablespoons rosemary
2 teaspoon ground cumin
2 teaspoon coriander

Combine the dry ingredients in the brine in a non-reactive pot with ½ gallon of water and bring to a simmer for 15 minutes. Allow the brine to cool (add ice cubes if necessary) and combine with the additional 1.5 gallons of water in a large pot that will accommodate the brine and the turkey.

If you are cooking for a large crowd, I recommend you purchase two small turkeys. They will cook in a shorter time period and will be younger (more tender) birds.

Submerge the turkey in the pot with the brine, weighting it down with a plate or pot lid to keep it totally submerged. You will need to clear space in the fridge or place the pot in a cooler surrounded by ice so that the brine temperature is 40°. Allow the turkey to brine overnight.

Heat your smoker to 350° prior to placing the turkey on the grill. Remove the turkey from the brine and pat it dry. Push your choice of herbs, dry rub, and butter under the skin over the breasts and legs as far as possible without tearing the skin.

Place the turkey, breast side up, on the smoker and cook at a temperature of 275° until the breasts register 160°. This should take 4 to 6 hours depending on the size of your turkey and environmental conditions. Don't rush it, there are games to watch and appetizers to taste.

Remove the turkey from the smoker and tent it with foil for 30 minutes before serving. I like to remove the breast in one solid piece with the skin attached and then slice it cross-wise in medallions so that everyone gets the full, flavor, including a taste of the skin.

Billy Bob Billy's
Smoked Duck

Impress your guests this holiday season and treat them to a sumptuous smoked duck cooked on your own backyard smoker or grill. Those of us on the competitive BBQ circuit covet secret recipes used in the "anything but" categories, which is anything but pork and can include fish, game or exotic fare. You may not want to jump into smoked ostrich or venison Oscar, but smoking a wild or farm raised duck will be an easy first step that is guaranteed to wow your guests.

Frozen, farm-raised ducklings can be found at most grocery stores and at a discount at Wal-Mart Supercenters. If you have a hunter in the family, wild duck has a more assertive (gamey) taste that many people prefer – some do not. Once the frozen ducks are thawed, the preparation, whether wild or farm-raised, is the same.

Marinating the ducks overnight in a brining solution or your favorite baste will add flavor and moisture to your duck. If your time does not permit marinating, don't despair, the low temperature smoking process will produce a tender, succulent bird.

Step one: Rinse the duck thoroughly under cold water. If farm-raised, remove the neck and duck parts that may be in the body cavity. If yours is a wild duck – check for any buckshot you can see or feel near the surface of the duck. If your farm-raised duck appears to have an excessive amount of fat – more than ¼ inch on the breasts – you may want to score the skin on the breasts in a cross-hatch pattern to allow some of the fat to render out in the cooking process. Pat dry and then apply your seasoning. Seasoning can be as simple as salt and pepper or your favorite secret dry rub you use for other cuts of meat. While it's not really necessary to *rub or massage* the seasoning into the duck, some chefs enjoy this process of bonding with their entrees. Do make sure you apply the dry rub liberally on all surface areas as well as inside the body cavity. Allow the dry rub to sit for 30 minutes before the next step.

Step Two: Place each duck into a plastic bag or non-reactive dish along with one cup of your marinade or basting sauce. Toss well so that your duck is coated and then place in the refrigerator overnight. Every couple of hours you should re-toss the bags to insure the duck is well coated with baste. Remove the ducks from the refrigerator one hour before cooking to allow them to reach room temperature, still in the bag, prior to cooking. This will avoid starting the cooking process with a cold center in your duck. Prior to placing your duck on the cooker, re-dust with your dry rub seasonings.

Step Three: Prepare your grill or smoker so that you have reached an initial temperature between 250° and 300°, because you will lose heat when you open the grill to add your ducks. If you are using a gas grill you can try to use only one burner. If the temp is still too high, your cooking time will be shorter and you will need to wrap your ducks in aluminum foil at the mid-way point to prevent drying out. For your charcoal grill, build your fire on one side and place you duck away from the fire. *See white man's fire!*

Soak your wood chips for at least one hour prior to use. Hickory works well, but if you can find apple or another fruit tree chips, the extra sweetness will be worth the effort. At the time you place the ducks on the cooker you should also place a handful of wet chips on your charcoal. The wet chips should be replenished whenever you see no smoke coming from the cooker. If you are using a gas grill, prepare several foil packets of wet chips with holes in the top of the packets. Place a packet near the burners to produce the smoke you need. Simply replace the old packet with a new one whenever the smoke is not longer visible.

The ducks should be basted every hour. While basting does provide some benefit to reducing the drying of the meat, the major benefit of using a flavorful baste is that you are adding layers of flavor to your duck. The hourly basting is a good time to check your charcoal and smoke chips.

Cooking time for a 6 to 8 pound duck is around three hours. If your cooker is maintaining 225°, four hours of smoking will not dry out your duck. If you are cooking wild ducks, because they are smaller and leaner, you may want to wrap them in aluminum foil with approx ½ cup of baste after two hours. This will rehydrate your ducks over the last hour. Remember, cooking on an outdoor smoker is as much art as science. Outside temperature, humidity, wind, etc. can have an effect on the cooking process. For a medium rare duck – *why would you spend all this time just to overcook your bird* – remove the duck from the cooker when the internal temperature of the breast is around 140°.

As an entrée, you should separate the breast and the leg, including the thigh, as an ample portion for any hungry guest. As an appetizer, I like to remove the breast meat and then slice across the breast in thin slices including the skin so that each bite includes the smoke flavor, the crisp duck skin, and a small amount of the duck fat just under the skin. The thin slices are then great when served on thin baguette toast or pita points with a dollop of cranberry sauce or heated apple butter.

Billy Bob Billy's
Duck Baste & Marinade

Ample for two farm-raised ducklings.

1 quart	Kraft Original BBQ Sauce	1 cup	Wickers marinade
1 quart	apple cider	1 cup	lemon juice

Plus the contents of one orange sauce packet contained in farm-raised duckling.

Combine all liquid ingredients, stirring in the BBQ sauce until dissolved.

This is great as an overnight marinade for chicken or ducks. Thin enough that it should not burn when used as baste, unless you are grilling at 400 degrees. Substitute your own BBQ sauce of choice.

Billy Bob Billy's
Duck Rub

2 tablespoons	Sweet Basil	1 tablespoons	Garlic Powder
2 tablespoons	Coarse Black Pepper	1 ½ teaspoons	Celery Seed
2 tablespoons	Paprika	1 teaspoon	Dry Mustard
1 tablespoons	Chile Powder	2 teaspoon	Ground Cumin
½ tablespoon	Red Cayenne Pepper	4 tablespoons	Brown Sugar

Billy Bob Billy's
Smoked Duck & Sausage Gumbo

1 Smoked Duck (see recipe)
1 pound Andouille sausage

4 teaspoons	salt	¾ cup	diced celery
2 teaspoons	duck rub (see recipe)	6 cups	chicken or duck stock
1 teaspoon	thyme	¾ cup	diced bell peppers
½ teaspoon	cracked black pepper		(red or yellow)
¼ teaspoon	cayenne pepper	1 bottle	stout beer
2	bay leaves		(Guinness will do)
2 tablespoons	garlic, minced	¼ cup	vegetable oil
1 ½ cups	diced onion	1 cup	all purpose flour

For serving:

5 cups	steamed white rice
1 cup	chopped green onions
½ cup	chopped parsley leaves

Debone all duck meat from smoked carcass. Set the meat aside. Cut the andouille sausage into ½ inch slices and brown in a separate skillet until crispy on both sides. Set the sausage aside.

Add the vegetable oil and flour to a heavy skillet to make a roux. Stir the roux continuously, using a wooden spoon, over medium heat until it reaches the color of dark chocolate. This may take 20 minutes or more. Do not rush it! If you burn the roux, you will have to throw it out and start over. *Nothing can save the bitter taste of burned roux.*

Add the vegetables and garlic to the roux and sweat until wilted. Add the liquid ingredients, the thyme, bay leaves, and duck rub, and stir to blend. Add the duck meat and sausage, then, bring the pot to a boil. Reduce the heat and simmer for 90 minutes to 2 hours. Occasionally stir the mixture to prevent burning.

Serve the gumbo over white rice and garnish with scallions and parsley.

(If you're really into this, create the duck stock by simmering the bones, wings, and skin with one cup each of onions, celery, and carrots for about one hour in 8 cups of water).

What's Your Game?

Wild game can be a real challenge since animals in the wild tend to be more muscled and leaner than the animals we raise for slaughter. In addition, their food sources are varied and can affect the taste of their meat. The majority of game harvested in this country is destined to be mixed with pork fat and ground into burgers prior to cooking. Game tenderloins can be easily prepared by grilling or pan frying without being turned into jerky, but larger cuts of meat may be a challenge. The key is to avoid overcooking the meat. Lean cuts of meat will dry out quickly once cut into pieces for your table, unless covered with a sauce.

Since venison is the most readily available game in this area of the county, I have included a recipe that has produced consistently good results.

Venison Roast

Citrus Cilantro Marinade

1/3 cup	lime juice
1/3 cup	orange juice
3 tablespoons	lemon juice
2 teaspoons	lemon zest
2 teaspoons	lime zest
½ cup	extra virgin olive oil
2/3 cup	chopped cilantro
4 cloves	garlic, pressed
½ teaspoon	cayenne pepper
½ teaspoon	coarse black pepper
¼ teaspoon	kosher salt
¼ teaspoon	oregano

Citrus Venison Rub

1 tablespoon	coriander seeds
1 tablespoon	orange zest
1 tablespoon	lime zest
2 teaspoons	black pepper
1 teaspoon	coarse salt
1 teaspoon	thyme

Marinade - Combine three juices and zest, stirring to mix. Drizzle in olive oil while stirring. Add spices, cilantro and pressed garlic for final stirring. (Can be made in food processor). Allow marinade flavors to marry for two hours in refrigerator before using.

Dry Rub - Toast coriander seeds in a dry skillet for 2 - 4 minutes. Crush with mortar and pestle into coarse powder. (Use back of wooden spoon in the skillet as substitute). Combine all ingredients in a bowl and whisk until uniform.

Debone venison and apply rub onto venison and allow to marinate one hour. Place venison in plastic bag with Citrus Cilantro Marinade for one hour or overnight in the refrigerator. One hour prior to cooking, remove venison from refrigerator. Dust with dry rub. Roll venison into a tight role and tie with baker's twine. Place in smoker at 225° for approx. three hours or until it reaches an internal temperature of 135°. Baste on the hour with a second batch of marinade. Remove roast and allow to rest 15 minutes. Serve hot, cut in one inch medallions or allow it to cool and slice thin for serving on toasted bread or for sandwiches. Also works well with venison tenderloin for a shorter cooking time.

Barbecued Orange Chicken

2 1/2 lbs chicken parts

BBQ Sauce:

1/4 cup	vegetable oil
1/4 cup	frozen orange juice concentrate
1/2 cup	white wine vinegar
1/4 cup	tomato paste

1 orange zest, removed with grater, orange slices reserved for salad

Salad:

1 large	ripe tomato
1	orange, sectioned (without rind, see above)
2	scallions, chopped
1 tablespoon	vegetable oil
1 tablespoon	white wine vinegar
1/8 teaspoon	salt
1/8 teaspoon	pepper
1 cup	apple wood chips

Soak apple chips for 30 minutes. Prepare grill for medium heat; heat coals and add chips. In medium bowl, mix together all barbecue ingredients until smooth. Place chicken on grill away from center heat, skin-side-down; cook 15 minutes. Turn chicken and grill for 10 additional minutes. Brush chicken pieces with sauce and turn occasionally; cooking for additional 10 minutes.

Cut tomato into wedges and place in medium bowl. Use sharp paring knife to cut out white pith off orange. Remove orange sections and add them to tomato. Sprinkle with oil, vinegar and scallions; toss. Season with salt and pepper and toss again. Serve chicken with salad on the side.

Maple Barbecued Chicken

4 skinless	chicken thighs	1 tablespoon	cider vinegar
3 tablespoons	maple syrup	1 tablespoon	canola oil
3 tablespoons	chili sauce	2 teaspoon	Dijon mustard

1 cup apple wood chips

Soak apple chips in water for 30 minutes. Preheat grill to medium heat. Combine syrup, chili sauce, vinegar and mustard together in a saucepan. Let simmer for 5 minutes.

Brush chicken with the oil and season with salt and pepper. Add wood chips to coals. Place chicken on grill and cook for 10-15 minutes or until fork tender. Turn occasionally and brush generously with sauce in the last few minutes before they are done.

BBQ Giant Prawns

12	Giant Prawns, shelled w/ tails	1 teaspoon	orange zest (grated)
1/4 cup	melted butter	2 each	green onions
1 cup	orange juice	1 teaspoon	ginger root
2 tablespoons	sherry		(freshly grated)

Soak a dozen long wooden skewers in water for 30 minutes. Then push skewers through prawns, lengthwise, from head to tail with only 1 to a skewer. Combine all ingredients in saucepan and cook over medium to low heat, stirring, until butter is completely melted. Dip skewered prawns in the orange sauce and position on oiled grill rack about 4 inches above the coals.

Baste liberally with sauce and grill for 2 minutes. Turn the prawn over and baste again, cooking for another 2 minutes. Smaller prawn will be done at this point, but continue basting and turning larger prawn until they are pink and cooked through. Remove from heat immediately when done, as they will get tough if overcooked. Use any remaining sauce for a dip for the prawns.

Bourbon Steak

1-½ inch cut of steak

1 teaspoon	sugar	¼ cup	water
¼ cup	bourbon	1 clove	garlic, crushed

1 cup	apple wood chips

Mix all ingredients together, place in plastic bag and marinate steak 4 hours. Soak wood chips in water for 30 minutes, and add to hot coals just before commencing to grill. Grill steak to desired doneness. This recipe is good with any cut of steak you like.

Marinated Steak Kabobs

1 cup	Onion, chopped	1 teaspoon	Dijon Mustard
1/2 cup	Vegetable oil	1 pound	Sirloin steak
1/2 cup	Lemon juice	1 large	Green bell pepper
1/4 cup	Soy sauce	2 medium	Onions, quartered
1 tablespoon	Worcestershire sauce	2 medium	Tomatoes, quartered

1 cup	apple wood chips

Cut steak in 2" cubes. Cut pepper in 2" squares. Sauté onion in oil; remove from heat. Stir in lemon juice, soy sauce, Worcestershire sauce, and mustard; pour over meat and vegetables. Cover and marinate overnight in refrigerator. Remove meat and vegetables from marinade, reserving marinade. Alternate meat and vegetables on skewers.

Soak apple wood chips in water for 30 minutes. Prepare fire in grill. When the grill is up to medium high temperature, add wood chips; let them start smoking. Grill kabobs 5 minutes on each side over coals or until desired degree of doneness, brushing frequently with marinade.

Billy Bob Billy's
BBQ Burritos

2 lbs. Pulled Pork
½ lbs Mexican chorizo sausage

1 large	yellow onion, chopped		¼	cup	apple cider vinegar
3	garlic cloves, minced		2		bay leaves
1 can	whole tomatoes, 16 oz		2		chipotle chilies, minced
1.5 cups	chicken stock		(use canned chipotles and reserve sauce)		

Salt and pepper to taste

1 package 10 inch tortillas

Remove chorizo from its casing. In a large skillet, sauté chorizo for 10 minutes breaking into small clumps. Remove chorizo, set aside, and sauté onions and garlic in sausage drippings in skillet until browned. Add shredded pork and sauté 5 – 7 minutes until pork is warmed throughout. Add tomatoes, stock, vinegar, bay leaves, chipotle chilies, and sautéed chorizo. Sauté for ten minutes, add the adobo sauce and sauté for 10 minutes more until mixture thickens.

Place tortillas on grill or high heat smoker toasting on each side for approx. 2 minutes until grill marks form on surface. Remove to plate, add several spoons of mixture, roll and enjoy with your favorite cold cerveza.

Cincinnati Style Smoked Chili w/Pasta

1 box whole grain linguine pasta

2 tablespoons	olive oil	3-4 cloves	garlic, crushed
1/4 cup	bacon bits, or 2 strips bacon, fried to crisp and crumbled	2 tablespoons	unsweetened cocoa powder
		2 tablespoons	chili powder
1 1/2 cups	onion, chopped	1/2 teaspoon	cinnamon
1 cup	green or red bell pepper, chopped (optional)	1/2 teaspoon	ground allspice
		1/2 teaspoon	ground cardamom
		1 teaspoon	cumin
1 (15 oz) can	chopped tomatoes	1/2 cup	red wine
1 (15 oz) can	tomato sauce	1/4 cup	wine or cider vinegar
4 ounces	tomato paste	2 tablespoons	honey or sugar
4	sun-dried tomatoes, minced or ground (optional)	A pinch crushed red pepper, to taste	

Serves / Yields
6 - 8 servings

Coarsely chop the pulled pork to bite sized pieces. Brown the onion and pepper in oil, until onion is transparent, then cool. Mix the rest of the ingredients and bring to a simmer. After 10 minutes, add the cooked meat, onions and pepper, then simmer for 30 minutes more. Cook the pasta until al dente, then drain. Serve the chili over the pasta and garnish with the other condiments as desired.

Helpful Hints: The meat may be cooked ahead of time, then refrigerated or even frozen before use. If spicy chili powder is used, the red pepper is not necessary. This is supposed to be a fairly mellow dish (not fiery hot). Whole cumin is preferable. Toast lightly first, then grind in a mortar or coffee/spice grinder. For that matter, use all whole spices when possible and grind just before using.

This recipe is best when prepared a day in advance, which allows the complex flavors to blend better.

Grilled Leek and Sweet Pepper Pasta

2 medium leeks, green tops trimmed, split up to the root ends, cleaned

1 tablespoon	olive oil	1 large clove	garlic, minced fine
1 large	red bell pepper	1/3 cup	dry vermouth
1 large	yellow bell pepper	salt and freshly ground pepper to taste	
4 tablespoon	unsalted butter		

12 oz.	fresh fettuccine
1 tablespoon	fresh thyme leaves
1 cup	apple wood chips

Prepare a medium-hot fire in the grill.

Trim the green tops off the leeks and slice in half lengthwise. Rinse under cold water to clean. Pat the leeks dry and coat with olive oil.

When the coals are covered in gray ash add the presoaked chips to the fire. When the chips start to smoke place the leeks and peppers on the cooking grill directly over the fire. Grill, turning as needed, until leeks are tender and golden brown - about 10-12 minutes, and skin of peppers is charred, about 15 minutes. Remove the leeks from grill and let cool. Remove the peppers from the grill and place in a paper or plastic bag and seal; set aside and allow to steam. When cool, trim root ends from leeks, then cut into thin strips. Peel and seed bell peppers and cut into thin strips.

Meanwhile, heat a large pot of water to boiling.

Heat butter in a large skillet over medium heat. Add garlic and cook, stirring frequently, until pale golden. Add vermouth, and reduce to syrupy consistency. Stir in leeks and peppers and season with salt and pepper.

Salt boiling water, add pasta, and cook until tender but still firm to the bite. Drain thoroughly, add to skillet and toss well. Sprinkle with thyme and serve hot.

Mediterranean Grilled Vegetables

1 pound	large onion	1 pound	zucchini squash
1 pound	red bell pepper	1 cup	olive oil
1 pound	green bell pepper	1/3 cup	Italian seasoning
1 pound	yellow squash		

(Kick it up a notch with Billy Bob's Basic Rub)
1 cup soaked apple wood chips

Peel onions and cut top-to-bottom in large wedges. Cut tops from bell peppers, remove core, and cut in large top-to-bottom pieces. Trim ends from squash and cut in diagonal rounds, about 1/2" thick. Toss all vegetables in a large bowl with olive oil and seasoning, breaking up the onion wedges somewhat.

Add soaked chips to the fire or in smoker box or pouch.

Place vegetables in a single layer on a very hot grill and cook, turning occasionally, until peppers are slightly charred and veggies are tender (about 5 minutes).

(Watch out for the flare-ups because of the olive oil dressing! Close the grill cover to reduce flare ups and infuse more smoke flavor)

Serve immediately. These are also good refrigerated then micro-waved to reheat.

Cinnamon Honey Wings

2 1/2 pounds   chicken wings

4 cloves	garlic, chopped	1 1/2 teaspoons	cinnamon, ground
1/4 cup	olive oil	1 teaspoon	thyme
2 tablespoons	soy sauce	1/2 teaspoon	ginger, ground
1/4 cup	vinegar, rice	1/2 teaspoon	mustard, dry
1/4 cup	honey		

| 1 cup | apple wood chips |

Mix all ingredients in a plastic bag then knead occasionally for 2 hours.

Soak wood chips in water for 30 minutes. Prepare fire in grill at medium heat. Add chips to hot coals. Cook wings on the grill for about 10 minutes on one side, then turn and baste with the marinade. Continue cooking 10 minutes or until done. Serve immediately or refrigerate until needed.

Grilled Halibut with Oriental Sauce

(4)   6-oz. halibut steaks

1/4 cup	orange juice	1 tablespoon	fresh lemon juice
2 tablespoons	soy sauce	1/2 teaspoon	oregano
2 tablespoons	ketchup	1/2 teaspoon	pepper
2 tablespoons	vegetable oil	1 clove	garlic, minced
2 tablespoons	fresh parsley		

1 cup   apple wood chips

Combine orange juice, soy sauce, ketchup, oil, parsley, lemon juice, oregano, pepper and garlic in a small bowl. Brush the mixture evenly on the steaks, refrigerate.

Brush the grill lightly with oil. Light the coals or gas grill. Soak the apple wood chips in water for about 45 minutes. When the coals turn white add the wood chips. When the chips start to smoke place the steaks on the grill rack and cook turning once, about 3 minutes per side, or until the steaks flake when tested with a fork.

Suzie Que's
BBQ Quesadillas

4	Whole grain tortillas	1 cup	Cilantro, chopped
	(corn or flour works, too)	1 cup	3 cheese Mexican mix
2 cups	pulled pork BBQ	2	jalapeno peppers

BBQ sauce of choice
Sour cream to taste

Pull leaves from fresh cilantro and rough chop approximately one cup. Thin slice two jalapeno peppers. To limit the heat, remove seeds and membranes before slicing. To kick it up a notch, substitute habanero pepper.

Place tortillas in a dry skillet until browned on both sides. Place one tortilla on your work surface and lightly brush with BBQ sauce. Spread ½ of the pulled pork then shredded cheese on the tortilla followed by jalapeno pepper slices and cilantro. Place second tortilla on top and return to the skillet for two minutes then turn and continue on other side for two minutes or until the cheese begins to melt. Repeat with last two tortillas and remaining ingredients. It also helps if you have a second skillet or weight to place on top of quesadilla while cooking. Serve with sour cream and salsa of your choice.

BBQ Quesadillas can be prepared on your grill if the grill surface supports the tortillas.

Salsa Recipes

Billy Bob Billy's
Smokin' Salsa

2	red bell peppers		3 tbs.	white wine vinegar
3	yellow tomatoes		3 tbs.	olive oil
1	jalapeno*		2 tbs.	parsley leaves
½	red onion		½ cup	cilantro leaves
2	garlic cloves		1	lime juiced

Salt & Pepper to taste
* kick it up a notch by adding ½ finely minced habanero pepper

Cut bell peppers and tomatoes in half. Remove ribs and seeds from peppers. Place peppers and tomatoes, cut side down, in smoker over direct heat with smoke. Smoke long enough to flavor, 5 – 8 minutes. While peppers are smoking, coarsely chop seeded jalapeno, red onion, parsley and cilantro. Remove peppers and tomatoes from smoker. Remove seeds from tomatoes. Coarsely chop peppers and tomatoes. Finely chop the garlic cloves. In a separate bowl, combine the garlic, vinegar, oil and lime juice. Wisk briefly. Combine all remaining ingredients in a large bowl, and drizzle in liquids while tossing. Salt & pepper to taste.

Billy Bob Billy's
Mango Black Bean Salsa

1	Mango, peeled and diced	1/3 cup	pineapple juice
¼	red bell pepper	¼ cup	cilantro
¼	green bell pepper	1 tablespoon	ground cumin
¼	red onion	½ tablespoon	minced habanero
½ cup	black beans	juice of two limes	

salt and pepper to taste

Peel mango and dice into small cubes slightly larger than pieces of chopped peppers. (You want them large enough that you get a burst of individual flavor when you bite down on the mango and bell peppers). Finely chop the red onion. Coarsely chop the cilantro.

Combine all ingredients and allow the flavors to marry covered in the refrigerator for one hour.

Note: Salsa takes on a different flavor the next day (it's all good), but if you make enough for two days or as a backup for your guests, then reserve the cilantro for one hour before serving. Cilantro tends to lose flavor over time or, even worse, can turn bitter overnight from interaction with the lime juice. Keep your salsa as fresh as possible for the best taste.

Mango Black Bean Salsa (hot)

1 large	mango	1/3 cup	pineapple juice
¼	red bell pepper	¼ cup	chopped cilantro
¼	green bell pepper	1 tablespoon	ground cumin
¼ medium	red onion	½ tablespoon	minced jalapeño*
½ cup	canned black beans	Juice of 2 limes	
Salt and pepper to taste			

*To kick it up a notch add ¼ tsp. minced habanero

Peel and dice the mango, dice the bell peppers. Coarsely chop the red onion.
Combine all ingredients in a bowl and enjoy.

I prefer to coarsely chop the onions and peppers so that you get a burst of individual flavors as you bite into the salsa. For a smoother salsa, all ingredients can be pulsed in a blender or food processor.

Billy Bob Billy's
Peach Salsa

2	fresh peaches		½ cup	cilantro
1	red bell pepper		juice of two limes	
1	yellow bell pepper		salt and pepper to taste	
1	jalapeno pepper		optional - sliced habanero	
½	red onion			

Peel peaches and dice into small cubes slightly larger than pieces of chopped peppers. (You want them large enough that you get a burst of individual flavor when you bite down on the peaches and bell peppers. Slightly firm peaches work best. I like to leave the skin on for added flavor and nutrients). Finely chop the red onion. Coarsely chop the cilantro.

Combine all ingredients and allow the flavors to marry covered in the refrigerator for one hour.

Suzie Que and I like to add a micro-julienned seeded habanero. Habaneros have a nutty flavor with intense heat, but if you cut them very thin, the peaches put out the fire very quickly after you bit into this salsa. **Hot enough to surprise you, but not so hot as to hurt anyone.**

Note: Salsa takes on a different flavor the next day (it's all good), but if you make enough for two days or as a backup for your guests, then reserve the cilantro for one hour before serving. Cilantro tends to lose flavor over time or, even worse, can turn bitter overnight from interaction with the limejuice. Keep your salsa as fresh as possible for the best taste.

Billy Bob Billy's
Cranberry-Pear Salsa

2	red pears (ripe, but firm)	1	lime
1 cup	sweetened dried cranberries (Cranraisins)	½	red onion
		¼ cup	chopped fresh cilantro
2	jalapenos	1 tsp	ground cumin

Wash pears and cut into small ¼ inch cubes. Coarse chop ½ red onion. Coarse chop cranraisins. Seed and remove ribs from jalapenos – fine chop. (You can kick it up a notch by micro-julienne slicing a seeded habanero in place of the jalapenos). Combine chopped ingredients and cumin in a bowl with the juice of one lime. Salt and pepper to taste. Combine all ingredients and allow the flavors to marry covered in the refrigerator for one hour.

Suzie Que and I like to add a micro-julienned seeded habanero. Habaneros have a nutty flavor with intense heat, but if you cut them very thin, the fruit puts out the fire very quickly after you bit into this salsa. **Hot enough to surprise you, but not so hot as to hurt anyone.**

Serve with chips as a traditional dip or puree in a blender and serve over smoked turkey for Thanksgiving dinner.

This salsa is great when served with turkey, but my friends always scarf it down with chips before the bird is ready. Suzie Que says it even goes down well with champagne!

Note: Salsa takes on a different flavor the next day (it's all good), but if you make enough for two days or as a backup for your guests, then reserve the cilantro for one hour before serving. Cilantro tends to lose flavor over time or, even worse, can turn bitter overnight from interaction with the limejuice. Keep your salsa as fresh as possible for the best taste.

BBQ Sides

Billy Bob Billy's
Baked Beans

2 16 oz	can pork and beans	2 tablespoon	ground mustard
2 16 oz	can ranch-style beans	¼ cup	brown sugar
1	green pepper, diced	¼ cup	BBQ sauce
1	medium onion, diced		(Kraft Original)
1 cup	shredded pork BBQ	½ teaspoon	celery salt
1 cup	sliced mushrooms	½ teaspoon	lemon pepper
¼ stick	unsalted butter	1 clove	chopped garlic
2 tablespoon	wine vinegar	Fresh basil and thyme to taste	
4 tablespoon	Worcestershire sauce		

Sauté onions, garlic and bell pepper in butter and 2 tablespoons of Worcestershire sauce until onions are translucent. Add mushrooms and 2 tablespoon Worcestershire and sauté until tender. In a large pot combine all ingredients. Bake at 350 degrees for 30 minutes.

For a great smoky taste: Grill halved onion and bell pepper until golden brown, dice and add to all ingredients in cast iron pot, then cook uncovered in the smoker at 250 degrees for 1 hour stirring frequently.

Serves 12 easily.

Billy Bob Billy's
BBQ Slaw

½ cup	apple cider vinegar	1 tablespoon	celery seed
½ cup	sugar	2 tablespoons	dry rub
½ cup	vegetable oil	2 tablespoons	Dijon mustard
6 medium	green onions	2 medium	cabbages

Take two cabbages (one green, one red, for color) and cut into quarters through the core so leaves will stay together. Brush the cut sides of the cabbages with vegetable oil and dust with dry rub (Billy Bob Billy's basic rub is a good choice).

Whisk the vinegar, sugar, celery seed, mustard, and vegetable oil in a non-reactive bowl. Season to taste with salt and pepper.

Grill cabbages and green onions on medium heat until grill marks appear. Approx. 4 minutes per side for the cabbages and 2 to 3 minutes for green onions. Transfer cabbages and onions to a cutting board and chop to desired thinness (discarding the cabbage cores).

Toss all ingredients in a large bowl to coat cabbage. Season the slaw with additional salt and pepper to suit your taste. Serve warm or chill for later serving.

This is a great warm weather slaw since there is no mayonnaise in the recipe. You may substitute your vinegar of choice. Susie Que likes to use red wine vinegar.

Great served "Memphis style" piled high on your pulled pork sandwich or served on the side with your ribs and BBQ beans.

Billy Bob Billy's
BBQ Bell Pepper Rings

Blaze and I spent many a Friday night at the Fox Tavern in Memphis. (It's probably out of business by now). One of our favorite sides to go with many a cold beer while we dominated the shuffle board was their fried bell pepper rings. The following stepped-up version is bound to please your friends.

3 – 4   Bell Peppers (mix green, yellow, red)

2 large   eggs
1 cup   all-purpose flour
1 cup   masa harina (Mexican flour)
1 bottle   Tecate beer

Your choice of dry rub.
Canola oil for frying

Slice your peppers in rings ¼ inch thick, removing the seeds, stems, and veins.

Combine eggs and one tablespoon of Billy Bob's Basic rub in a bowl and whisk vigorously. In a shallow dish, combine flour and two tablespoons of dry rub. Mix thoroughly. In another bowl, combine masa harina and beer. Whisk gently until smooth.

Place a deep skillet or Dutch oven over hot coals with enough canola oil to come 2 inches up the side. Don't overfill – This is very flammable over a hot fire! Heat the oil to 350°.

When the oil reaches 350°, take the peppers, one at a time and season. First, dredge the rings through the flour and rub mixture. Shake off the loose flour. Dip the rings in the eggs and allow the excess to drip off. Then, dredge through the beer batter and allow excess to drip off before placing in oil. Place the rings in the 350° oil one at a time, being careful not to overlap the rings. *They will stick together. Don't overload the oil or yo*ur temperature will drop and your rings will be soggy. Cook the rings until brown on both sides, about 2 minutes per side. Remove the rings and dry on a paper towel or paper bag. Dust with additional dry rub while hot.

For a thinner, non-alcoholic version, simply use the dry masa harina for the second coating. This makes a crispy ring and the juiciness of the bell peppers really comes through.

BBQ Chicken
Buffalo Wing Dip

2	BBQ'ed chicken breasts	8 ounces	blue cheese, crumbled
8 ounces	Neufchatel Cheese	12 ounces	wing sauce
8 ounces	ranch dressing	1 cup	celery, chopped

Tabasco sauce or your choice, to taste.

Smoke two chicken breasts (or one whole chicken), using the recipe of your choice. Pull the breasts and shred with forks, then coarsely chop. You want the chicken pieces small enough to be lifted by your choice of chips or pita points. Slice celery in thin slices across the stalk to yield one cup (approx. two stalks). Allow the cheese to soften at room temperature prior to mixing. You can use cream cheese, but the Neufchatel has 40% less fat – not that you care! Combine the soften cheese and the ranch dressing in a bowl, stirring until smooth. Add the wing sauce and hot sauce while stirring to achieve your desired level of heat. Add chicken and celery, stirring until fully blended.

Pour into 9 X 9 inch baking dish and heat in pre-heated oven at 350° for 15 to 20 minutes until the surface is bubbly. The dip does not need to cook – only heated thru. Remove from the oven and allow to cool enough to be eaten. Sprinkle with crumbled blue cheese before serving.

Make this dip Volcanic by adding a sliced habanero prior to heating or simply add the habanero prior to serving to add a surprise to some bites. If the peppers are added prior to heating, the heat will spread through all of the dip!

Sunrise to Sunset

A Bloody Mary is always a good way to start a day of smoking. We all know a large amount of beer may be consumed during the day. So, this is a good way to jump start your day and get your first serving of vegetables. Suzie Que and I disagree on who's recipes this is, so I am also listing her version.

Billy Bob Billy's
Bloody Mary

2 cups Major Peters Bloody Mary mix
2 cups Clamato Juice
¼ cup Worcestershire
¾ tablespoon ground horseradish
2 tablespoon dry rub

1 tablespoon lime juice
1 teaspoon hot sauce
2 pinches coarse salt
Ground pepper to taste

Combine ingredients in large serving pitcher.
Rim glass with dry rub, add ice, Bloody Mary and your favorite garnish

Suzie Que's
Eye Opener

2 cups Major Peters Bloody Mary mix
2 cups Clamato Juice
¼ cup Worcestershire
¾ tablespoon ground horseradish
2 tablespoon dry rub

1 tablespoon lime juice
1 teaspoon Tobasco sauce
2 pinches coarse salt
Ground pepper to taste

Combine ingredients in large serving pitcher.
Rim glass with dry rub, add ice, Bloody Mary and your favorite garnish
Suzie Que likes to micro-julienne a fresh Habanero pepper and add some to her glass.

Reidarita

3 shots Cuervo Gold
2 shots Duffey's Lime Juice

1 shot Martinelli's apple juice
1 shot triple sec.

Shake 'til cold…serve wet!

This recipe comes from son-in-law Reid Jackson in Denver, CO. It's guaranteed to give you a Rocky Mountain High. It's a great way to end the day, but treat it with respect. Goodnight!

__Libation Log__

Competition

Competition is a natural component of the human condition. Whether you are challenged to improve your cooking prowess or trying to better your neighbor's or a family member's last bar-b-que production for bragging rights, it's natural to want to measure your success.

Memphis in May (MIM) is the granddaddy of BBQ competition. You can easily drink a case of beer arguing with a Kansa City BBQ enthusiast about which is best - the MIM World Championship Barbecue Cooking Contest format or the Kansas City BBQ Society - Kansas City American Royal Contest. Covington, TN claims to have the oldest and longest-running BBQ contest in the nation, but I'd group it in the back-yard-political category. In addition, many states and regions across the USA and Europe are now forming their own sanctioning organizations and holding exciting and fun contests. Most contests follow either the MIM (now Memphis BBQ Network) format or the KCBS – Kansas City BBQ Society format, or a slight variation of one of the two.

In the beginning . . .
In 1978, a group of BBQ enthusiasts gathered under a tent in an empty lot behind the Orpheum theatre for the first BBQ contest in Memphis. Twenty teams cooking under a tent proved to be a mistake in judgment, but fortunately the BBQ learning curve is very forgiving and competitors are a hardy bunch. A motley crew of Webers, barrels and wash tubs saluted the first winner, Bessie Lou Cathey and her "family famous" ribs. She won $500 for her efforts after paying a $12 entry fee. Pork shoulder and whole hog were added the next year, although MIM records on those two categories are incomplete earlier than 1982. The current MIM World Championship limits the number of competitive teams to 250 and the entry fee has gone up a little – by a factor of 100!

Since I cut my teeth on the Memphis-in-May style contest, it is the contest format I prefer. All MIM categories - ribs, shoulder and whole hog, are obviously pork. Preliminary judging takes place on-site and includes the team's presentation to three individual judges. Blind samples are also sent to a judging tent to be evaluated by four judges seated at one table. The top three teams in each category are then judged by a team of four finals judges who visit the nine teams on-site for a presentation and sampling of their product. I like this aspect of the MIM contest because it's a winner-take-all format. You have to be the be the best-of-the-best and when the judging is right, which is more often than not thanks to judges training schools, no cut of meat is favored over another. *Historically, ribs seem to have a slight edge, followed by whole hog, then shoulders.*

Kansas City BBQ Society contests require competitors to cook four categories of meat – chicken, pork ribs, pulled pork (shoulder or Boston butt), and beef brisket. All judging is done in a centralized judging tent in the "blind" form. The team with the highest point total including all categories is the Grand Champion. There is no on-site judging.

KCBS competitors tell me MIM contests are too hard because of the on-site judging and there's too much "politikin" going on during the presentations. It's true, that if you are competing in all three categories you will have nine on-site presentations as well as blind boxes of pork to submit. And, if you are lucky, you may also have three presentations to a group of finals judges if you are in the top three of all three categories. But as far as on-site politicking is concerned – contests are won in the blind judging, not the on-site dog and pony shows.

MIM competitors tell me that they think it's difficult to cook four different products, often on one smoker. Each cut of meat may need different time, temperature and basting. MIM competitors, me included, find it frustrating when the top cookers in the four categories are all beat by a team who may score consistent fourth place scores in all categories. (This team would not even make the finals at MIM since they were not in the top three in any category.) I'll have to admit that the KCBS format gives more teams a chance to win against the "professional" teams that are competing with every weekend.

A MIM competitor has the opportunity to receive feedback over and above their scores from on-site judges after judging has been completed. (See the section on scoring). This can go a long way to improving your cooking and your scores.

The MIM World Championship has now grown to a week-long celebration resembling a riverfront Mardi Gras. It has become very expensive to compete in the "big dance" in Memphis, but there are many, many local contests using the MIM and KCBS contest format that anyone can enter and have a good time competing against great members of the "BBQ Nation."

Changing trends in BBQ

As more teams enter competition, the preferences in BBQ flavor have gone through a number of changes in part because of the increased sophistication of competitors, but in a large part because of the increased involvement of competitors and judges from different areas of the country with different tastes.

It's long been acknowledged that western barbecue was primarily beef with a spicy red sauce. In the south a sweeter tomato based sauce is preferred, except for Chris Lilly and his "Alabama" White Sauce that has carried the Big Bob Gibson BBQ team to many a championship. In the Carolinas, vinegar sauces are the choice in North Carolina, while South Carolina likes their mustard based sauces. That's not the element of taste that I'm referring to.

When the Holy Smokers Too team was kicking butt on a regular basis, barbecued pork was in its purest form. Slow cooked barbecue with a wonderful smoky flavor imparted by cooking with a regional hardwood, usually hickory. A light vinegar baste was often used, but the end product was not flavored by anything other than hickory smoke, fire, and the internal juices. BBQ sauces were generally served on the side and called a "serving sauce" to differentiate it from basting sauces.

In 1990, the Apple City Barbecue team from Murphreysboro, Ill. came onto the competition scene and a major change appeared in many team's cooking process. The Apple City gang used both apple wood, apple cider and apple juice in their cooking process. Many teams were using fruit juices before 1990, but when this team took the contest scene by storm, people took notice. These guys were the first team to when MIM three times, and ended our run as the team to beat.

After their third win, Mike Mills headed to Las Vegas and opened the first of his successful Memphis BBQ restaurants. Pat Burke and the gang continues to compete as the Tower Rock team – winning the whole hog category at MIM in 1998.

Following the shift to apple products in the cooking process, the trend toward sweet continues. Today, I judge many successful teams that are using large quantities of brown sugar and/or fruit preserves in their cooking process – many to an extreme. If cooked at a high temperature, this can yield a product with a blackened, caramelized surface. It's not my favorite technique, but has gained favor to satisfy this country's sweet tooth.

Myron Mixon of Jack's Old South produces a sweet-coated product. I've judged Myron when the sugar barq' has caramelized to a crunchy coat and I'm sure my score was thrown out. Since Jack's Old South is the most successful team in the history of barbecue contests, I guess my opinion is in the minority. Oh well, it is *my* opinion. When Myron's team is "on que", they produce an end-product that in 99 out of 100 contests will take home the title. He has been so successful that he sold his restaurant and is now a full-time competitor as well as a producer of barbecue products and cooking seminars.

Recent contests have shown success for the "grilled finish" style of rib cooking. Competitors will smoke their ribs in the traditional style for wet or dry ribs. In the last phase the ribs are sauced and placed on a hot grill to "finish-off" the ribs. This produces a crispy caramelized coating of sauce. This is standard procedure for the contestants at the Reno Rib Cook-off where all competitors prepare St. Louis style ribs for hungry attendees during a week-long marathon of porcine professionalism. While this is new to the southeastern Memphis BBQ Network competitors, it's really not new at all. The world-famous Rendezvous restaurant in Memphis has been grilling their ribs and dusting them with dry rub just before serving for years. Dreamland BBQ in Georgia and Alabama grill their spare ribs for 90 minutes before slathering them with sauces and serving them up.

If you are going to compete and win, you better take notice of changing public opinion. America's sweet tooth is taking over BBQ taste preferences. Just remember that the judges around the circuit come from the general population and they prefer sweet. It doesn't matter if your garlic rubbed ribs with habanero sauce light up your neighborhood on a regular basis. Their wonderfully "unique" flavor doesn't have a snowball's chance in hell of pleasing the ten judges that will compare your ribs to the sweet offerings of your competitors.

I don't like the sweet trend. I prefer the Memphis-style dry ribs, but again that's just my opinion. Fortunately, since I no longer compete, I can cook what I like to eat and focus on pleasing me, Suzie Que and my friends.

Holy Smokers Too's Decade of Dominance

Every dog has his day and ours was from 1985 to 1994. During that period of time we accumulated too many trophies to carry in our old motor home. We adopted a policy of only carrying our MIM trophies and those that pertained to the contest we were competing in that weekend. During those ten years we were fortunate to make it into the finals at the World Championship five times, win the whole hog category four times, and win the Grand Championship twice. *The 1985 Holy Smokers Too team was one of a select group of teams sent to Ireland to compete in a truly international contest. That team came in second in the Irish International Contest!* Back then it was called the Memphis in May International BBQ Fest. The world championship trophies back then were modest wooden trophies with a small brass plaque not as large as neighborhood soccer trophies. After being hauled from contest to contest the only remnants from our world championship trophies are the brass plaques I mounted on stained cutting boards.

But, one reward that is of great pride and has not fallen apart is our world championship rings. I mention this because the background story is very interesting. As I have said, this is a very competitive sport. Many a winning team's egos have outpaced their BBQ. In 1983 and 1984, John Willingham's River City Rooters won back-to-back world championships for their spare ribs. John was so sure he was going to win that he had Memphis in May contact a local jeweler to design a championship ring using the MIM logo. He has never been thrilled that Holy Smokers Too took possession of the mold in 1985 and he has never won again. We like to hang our hands over John's fence at contests when we drop by to chat!

In 1994 we met our Waterloo. Pat Burke and Mike Mill's Apple City BBQ had taken the competitive circuit by storm. Their use of apple wood and apple juice was producing a rib that few teams could beat. In 1994 we found ourselves standing on the stage at Memphis in May as the winners in the whole hog category looking across the stage at our good friends from Murphysboro, Ill. who were the winners in the rib category. It was not our best hog and even if it had been, would not have been able to beat a great rib. We were glad that they were the first to win a third championship if we could not be the ones.

But, later, the thought of trying to win twice to take the lead seemed to be too much of a task. We had already lost some of the old guys who helped win in 1985. The new members we brought on for a needed infusion of youthful energy were more interested in partying than the hard work it takes to prepare for and clean up after a contest. We had too many type-A personalities who wanted to try new techniques. That yielded Alabama ribs so hot the judges ran down the road like Richard Pryor on fire or Australian ribs so salty even the flies would not land on them. In the mean time our lack of focus diverted us from the original whole hog process that had served us so well for 15 years. The number of annual contests dwindled from 12 to 4 – 6. I and a team member in Missouri found it hard to be present for the work details. In 2006, the 25th year of competition for the Holy Smokers Too, I called it quits and retired from competition. I had been judging contests and teaching BBQ cooking classes at the University of Tennessee for about five years. That was beginning to be more fun and more rewarding than the competition circuit, given our team's lack of focus. It was time for me to go.

I still compete, as a mentor, with friends I have met along the way or some of my students. And cook with my daughter and son-in-law whenever the Holy Smokers Too Rocky Mountain High team competes. They seem to enjoy the Denver lifestyle more than the competition. If you're going to hang out smoking BBQ for the weekend, *why not do it with good friends*? They, too, party so much they lose track of the proceedings. They missed the awards ceremony in their first contest in Dillon, Co. (The highest BBQ contest in the USA) and did not find out until two weeks later that they had won the salsa competition. Yep, they are true members of the Billy Bob Billy BBQ family!

You'll never find a friendlier group of people than those at a BBQ contest. I think the best are at the MIM format competitions because they are naturally set up to invite visitors into their area just as they would a judge. Many a team has eagerly lent a hand to a competitor who forgot a necessary item or just need help lifting their hog. If you're not ready to enter a contest, don't be shy about introducing yourself to a team and asking them about their BBQ. Just remember, if they are in the middle of preparing for judging or feeding a group of friends, they may ask you to come back later. But, rest assured that the number two activity behind smoking BBQ is talking about smoking BBQ. *A cold beverage goes hand-in-hand with either activity.*

For those of you just beginning to think about entering a contest, there is a backyard division at most contests – MIM calls it the Patio Division. These are for BBQ enthusiasts who don't have the experience or equipment to compete with the big boys. Most contests require you to move up to the big contest if you win the backyard contest. Many winning backyard teams change their name for the next year to avoid the expenses and truly feel they have more fun in the backyard. My friend, J.J. Butz in Charlotte claims that is why he has never won the backyard contest at the Blues, Brews & BBQ contest held in September. Of course, that does not explain why he hasn't quite made it to the turn-in stage in some prior contests. *Who says you can't have too much fun?* The back yard contest is a great way to dip your toe into the contest scene. BB & B in Charlotte is a great beginner contest because Smithfield has been providing the free meat for past contestants.

Another good starting point is to attend a BBQ cooking school that includes contest rules and tips. Several competitors offer classes either independently, affiliated with a contest, or aligned with a school such as my classes at the University of Tennessee. You can search online and find one close to your area or come to my class. **www.outreach.utk.edu/ppd or www.holysmokerstoo.com** .

Once you have read through the chapter on competition rules and techniques, I believe you will be well on your way to a successful new hobby, but a little hands-on training via a class or workshop can only help your progress towards taking home a few trophies and bragging rights for yourself.

Memphis in May Contest Rules

Barbecue is defined by the Memphis in May Championship Barbecue Cooking Contest as pork meat. The meat must be fresh and uncured and cooked on a wood fire (charcoal or natural wood). Gas or electrical starters may be used, as well as flammables to start your fire. Once the meat is placed in your cooker, only wood products can be used for heat.

Meat is inspected during the day prior to the contest (Friday). Meat preparation cannot proceed until after inspection. You can trim your cuts of meat, but you cannot begin applying rubs, baste or injections, until after inspection is completed.

Meat Categories

Whole hog – whole hogs that have been dressed by a certified slaughter house must weigh at least 85 pounds. The entry must be cooked as one complete unit, but removal of the head, feet or skin is permissible. The hog must remain whole during the entire cooking process. *No fair separating the ribs, shoulders or hams and then putting them back together for judging! Our good friends with Porky Pilots won that way one year and MIM quickly closed that loophole.* The submitted entry for judging must include portions of the shoulder, loin, and ham at a minimum. Other additional portions may be submitted if the team chooses.

Pork Shoulder – MIM defines pork shoulder as containing the arm bone, shank bone, and a portion of the blade bone. Hams submitted must contain the hind leg bone. *Boston butts or picnic hams are not allowed. Smaller contests will usually allow Boston butts.*

Pork Ribs – Either pork spare ribs or loin (baby back) ribs are required. *St. Louis cuts would qualify, but baby backs have proven to be the preferred cut with judges.*

Patio Porkers and Backyard competitors are generally limited to ribs or Boston butts because of the size of their equipment. Miscellaneous–Competitive categories have grown to include a wide variety of contest offerings. Everything from the "Anything Butt" competition that includes any non-pork offering you can cook on a smoker or grill. For example: Beef, fowl, fish, and exotics such as Kangaroo or 'gator. Add to that desserts, sauces, rubs, side dishes, T-shirts, hats, booths, etc. and you can find a way to display your unique creativity.

I actually competed in a contest in my adopted home town of Knoxville where the prize for the bean eating contest was greater than one of the pork categories! Needless to say, not many MIM teams were interested in coming back to that contest.

Judging & Scoring

Judging for Memphis in May contests includes both an on-site presentation component and a blind judging component. Three on-site judges will arrive at competitors' areas on twenty minute intervals and will judge your product for 15 minutes of that period. *You have five minutes between judges to clear the deck and prepare for your next judge. Should one judge put you behind schedule, you have the right to ask the next judge to give you additional time to prepare.*

On-site judges will evaluate your team and your product in five categories as well as an overall impression.

Area and Personal Appearance (2%) – Your team and your area will be judged on its appearance. Have you presented a neat, clean and attractive area? *Did you clean up after last night's party?* The amount of money spent or side beverages served is not a consideration.

Presentation (4%) – You will be evaluated on how thoroughly and clearly you have explained your cooking process. This may include: team origins, your unique cooking method, design of your cooker; development of sauces, rubs and marinades: presentation of your entry as well as how you took your entry from its raw state to the finished product. *Humor or a slight embellishment is not discounted!*

Appearance of Entry (24%) – Does your entry look appealing? Is it something you would want to put in your mouth? It should look good in its final stage on your cooker as well as on the plate or in the blind box. *Remember, we first eat with our eyes!*

Tenderness of Entry (29%) – Tenderness is a relative measurement, but one of the most heavily counted scores. Your finish product should have a firm (not mushy) texture, but contain moisture and pull easily from the bone. It should be easy to chew. Dry and stringy just won't do.

Flavor of Entry (24%) – Does the barbecue have a good taste? Most judges will first taste the meat by itself and then taste it with the sauce (if you provided it). Your sauce should compliment, not overpower, the meat. *If you submit your entry without sauce on the side or with sauce in the meat, the judges are not supposed to discount your score. But, remember, this is tough competition and everyone has different tastes. You want to give your entry its best shot at a win.*

Overall Impression (18%) – Overall impression is not an average of your scores, but it is a ranking of the judge's total experience judging your team. This is also where a judge has the ability to separate two excellent teams. This score can be given in tenths to break ties if the judge has two teams that score tens in all categories.

Blind judging is conducted in a private area where the products of 4 – 5 teams are judged at a table of four judges. Blind judges do not score entries in the first two categories of Area and Presentation. The lowest score in each of the six categories is dropped to try and weed out the outlier score.

The percentages in parenthesis are the relative importance of each category based on my calculations of the scoring process from past contests. Because of the implied multipliers that give more weight to a ten in Flavor than a ten in Area, you can see which categories are the most important. At the peak of performance in the most competitive contests such as the MIM World Championship, it's doubtful that any team with more than one nine in the categories of tenderness or flavor will make the finals. Just look at how close the scores were in the whole hog category for the 2002. Only 3 points out of 1020 separated the three hogs going into the finals.

Some teams make the mistake of pulling their best meat and holding it for the finals. *I call that take home meat!* If you do not submit your best product for preliminary judging, you will not make the finals and that saved meat will be going home with you. Remember, the meat you submit in the blind box has to stand on its own. You won't be there to tell the judges how great it is.

Now that the Memphis Barbecue Network has taken over for MIM in sanctioning and judging contests, the scoring may change a little. Maybe not! Regardless, I don't think the relative importance of the categories will change. Flavor and tenderness will always be the most important factors in evaluating barbecue.

HOLY SMOKERS TOO - CONTEST SCORES

MIM World Championship Barbeque Cooking Contest 2002

Preliminary	J1	J2	J3	J4	J5	J6	J7	Totals	Multiplier	Score	
Area	10*	10	10					20	1	20	
Presentation	10	9*	10					20	2	40	
Appearance	9*	9	10	10	10	9	10	58	4	232	
Tenderness	10	9*	10	10	9	9	10	58	5	290	
Flavor	9	9	10	10	10	8*	10	58	4	232	
Overall	9.6*	9.7	10	9.9	9.8	9.7	10	59.1	3	177.3	
Total	29	37.7	60	39.9	38.8	27.7	40	273.1		991.3	

MIM Perfect Score Model

Perfect Score	J1	J2	J3	J4	J5	J6	J7	Totals	Multiplier	Score	
Area	*	10	10					20	1	20	2%
Presentation	10	*	10					20	2	40	4%
Appearance	*	10	10	10	10	10	10	60	4	240	24%
Tenderness	10	*	10	10	10	10	10	60	5	300	29%
Flavor	10	10	10	10	10	*	10	60	4	240	24%
Overall	*	10	10	10	10	10	10	60	3	180	18%
Total	30	40	60	40	40	30	40	280		1020	

* Dropped low criteria score

Bearfoot BBQ in the Smokies 2002

Preliminary	J1	J2	J3	J4	J5	J6	J7	Totals	Multiplier	Score	
Area	10*	10	10					20	1	20	
Presentation	10*	10	10					20	2	40	
Appearance	10	10	10	9	8*	9	10	58	4	232	
Tenderness	9	9	10	8*	9	10	9	56	5	280	
Flavor	10	10	10	9*	9	9	9	57	4	228	
Overall	9.9	9.8	10	9.0*	9.5	9.3	9.3	57.8	3	173.4	
Total	38.9	58.8	60	9	8	37.3	37.3	268.8		973.4	

Memphis in May scoring format consists of three on-site judges (J1 – J3) and four blind judges (J4 – J7). The lowest score in each category is dropped as indicated by an asterisk next to the number on the score sheet. The top and bottom above are the actual scores received by Holy Smokers Too at Memphis in May and a contest in Knoxville, TN. The middle scores represent a perfect score in all categories.

You will notice a multiplier in the column before the final scores. The MIM system weights scores by this multiplier to give greater value to a score of ten in flavor than a score of ten in appearance or presentation. This multiplier is a closely guarded secret in the MIM and now MBN system. The multipliers and therefore the highest possible final score have changed slightly over the years, but this will give you a good idea of the importance of each judging category. The most important category, and greatest challenge, is tenderness which accounts for almost 30% of the total score. Appearance and flavor run a close second at almost 25% each. As you will see on the following team score sheet, more than two nines in the prime categories or a blown presentation can keep a team out of the finals.

I came by this discovery of the multipliers by utilizing the massive computer power of the Oak Ridge Super-computers to solve the formulas using higher order quadratic equations. We were able to approximate the multipliers but never quite reached a final determination. Then one night after a case of beer and a bottle of vodka tonics, I was able to coax the final clues from a former MIM official who had been involved with the development of the original computerized scoring system. We will just call him "Mystery Cannon" or MC because he fired the shot heard 'round the competitive world. ***Now that it's in my cookbook, that is!*** After several years of coaxing and plying MC with alcohol and then my n'th iteration of applying random numbers (I was just guessing), I was able to duplicate our final scores using these multipliers.

Rumor has it that you may see MC hanging around with the Parrot Head Porkers at some contests. If you sneak by late at night it's a good bet you may see him, but it's doubtful he will be able to see you.

The accompanying whole hog team scores are also from the 2002 MIM contest. I list them just to show you how competitive the contest is. It's not unusual to have only tenths of a point separating the top three teams out of a possible score of 1020! These three whole hog teams went into the finals with a clean score sheet against the top three rib and shoulder teams. It doesn't matter how good your preliminary scores are, your product and your scores in the finals have to stand for themselves.

Some of us old competitors have stated that the contest does not really start until the finals. With the challenge of 250 competitors and at least 250 judges of different levels of experience, the preliminary round is sometimes called "the lottery." If you win the lottery, you get to enter the nine-team finals contest.

In the early years, the scores were even tighter than today. Judges started every team with a perfect ten in each category and then reduced the scores if they found a flaw. It was possible to have two perfect teams scored by one judge with the only differential being the score for overall impression that can be scored to a decimal point. While I agree with this method, I can see where you might end up with a number of tie scores. The judges are now instructed to judge the product against the other teams they score. In other words, a judge that judges three teams can only give one ten for flavor, because two teams can't have the "best" you judged today. As you can see, only one of three on-site judged teams will receive a ten from one judge - in the blind, only one in four or six. Like basketball tournament brackets, if the two top teams are judged by the same judges, only one will likely make the finals. I have to assume that judge assignments take this into consideration so that the same three on-site judges do not judge the same three teams. But, in the blind (where contests are won) this is not possible. You may have four judges judging up to six entries. If you follow the guidelines in a strict sense, you'll have to give the teams flavor scores of 10,9,8,7,6,5. *(You get a 5 just for showing up, even if you don't submit meat for judging.)* In actuality, MIM representatives check turned in score cards to see that you did not give tens to more than one team in a category and the overall impression scores are not the same so it can serve as the tie-breaker. In a perfect tie, the judges will go back to the score card that was thrown out and add it to the mix.

Using this scoring system it is possible to increase the chance that a perennial favorite makes it to the finals by putting that team in a judging rotation that only includes first time local competitors. On the other hand, you could put all the "big boys" on one blind table and give the local boys a better chance grouped on a different table. The best of four bad entries on one blind table should get tens! Of course, if only local competitors win at your contest, it's not likely the "big boys" will come back next year.

I have no direct knowledge that any such gaming of the system takes place. And looking at judging alternatives, I have not seen a better process for delivering experienced judging and a quality result. I would like for all finals judges to be put thru a contest, working with a real competition team before they become certified to be a finals judge. It's discouraging to a team that has invested considerable time and money to travel to a contest to compete when a local celebrity, who doesn't have a clue about great BBQ, is a finals judge.

One of the very positive byproducts of the on-site judging versus a total blind judging, as in the KCBS network, is feedback to the team. A team is not allowed to give gifts to on-site judges while judging –"Look for the $100 bill under the plate, sir." But, you can ask your judges to stop back by after judging for BBQ or a token gift. This is a great opportunity to ask for feedback. Most of you returning judges may be the ones who thought your product was great so you won't get much constructive criticism, but if you ask what you could have done better, you may gather a few tips to help you improve. Otherwise, how will you know what areas need work? If the judge thinks you were perfect, ask him what types of errors did he find with other teams? Remember, if he gave you a ten, somebody had to get an eight or nine. If you're a new team or know your Que is not the best you can do, admit it to the judge and ask him or her to come back later and give you input. It will not negatively impact your scores and may even increase your overall impression score – get those points where you can!

Your goal should be to have a great time and improve your cooking skills. Trophies, plaques and ribbons are just a bonus in recognition of your efforts and success on that particular day.

Besides, we found it to be quite a storage and maintenance problem once we collected over one hundred trophies!! We only have the brass plates from the front of our 1985 and 1988 MIM Championships. The original wooden trophies have long since been destroyed from being carried from contest to contest. The new metal trophies hold up better, with the occasional loss of an ornamental pig or meat cleaver.

As you can see by the Memphis in May team score sheet from the 2002 World Championship, there were fewer than three points separating the top three teams who made it into the finals and only 14 points down to the top ten. More than one nine in appearance, tenderness or flavor and you did not make it into the finals in 2002. The top ten had no more than three nines in the key areas…that's a pretty tight contest. Jack's Old South in fifth place was only one nine-point score out of first place!

In 2002, Southern Gentleman's Culinary Society went on to win the whole hog category in the finals, but the Pyropigmaniacs team went on to win their second Grand Championship with their kick-ass ribs. Way to go George!

5/19/2002

PLACE	AREA	TEAM NAME	SCORE	
1	134	PIT & PIGULUN II, THE	1016.2	FINALIST
2	142	SOUTHERN GENTLEMEN'S CULINARY SOCIETY	1015.7	FINALIST
3	163	PIG POUNDA KAPPA	1013.9	FINALIST
4	145	CURLY TAIL SMOKERS	1011.9	
5	138	JACK'S OLD SOUTH	1009.8	
6	140	SUPER SMOKERS BBQ (ST. LOUIS)	1009.1	
7	149	AIRPORK CREW	1005.2	
8	161	HUFF-N-PUFF PORKERS	1004.1	
9	135	SMOKIN' STOKES & CHEERWINE	1003.1	
10	157	BARBECUE REPUBLIC	1002.8	
11	129	CUSTOM COOKERS	1002.7	
12	133	CRISPY CRITTERS	999.8	
13	162	SMARR COOKING CREW	998.7	
14	137	FLORIDA BOYS	998.4	
15	136	VIENNA VOLUNTEER FIRE DEPARTMENT	997.9	
16	148	SUPER CLEAN SMOKERS	997.6	
17	154	HOLY SMOKERS TOO	991.3	
18	127	GENUSWINE	987.9	
19	141	PRIME TIME COOKERS	981.1	
20	144	SUPER SWINE SIZZLERS	980.7	
21	156	GOOD, THE BAD AND THE SWINE, THE	979.9	
22	150	PIT-STOP PORKERS	976.5	
23	160	KING'S COOKING	974.9	
24	125	HOGAPALOOSA	972.6	
25	164	PIG DUAMONDS	972.4	
26	143	AOC HOGS	970	
27	130	WILLINGHAM'S RIVER CITY ROOTERS	966	
28	132	PORKY PILOTS	960.8	
29	139	ANNESDALE PORK	958.9	
30	166	DIXIE CREEK RIB TICKLERS	958.1	
31	155	JOWLS	952.8	
32	147	BAR-B-QUE BLUFFERS	952.5	
33	153	SOUTHERN PRIDE WHOLE HOG COOKERS	947.6	
34	165	HARVEST HOGS	945.1	
35	167	BADGES, BREWS & BBQ'S	945	
36	158	MIGHTY SWINE, THE	943.2	
37	128	REDNECK BAR B Q EXPRESS	942.3	
38	152	ELVIS PORKSLEY'S GREASELAND PORKERS	939.1	
39	159	TEAM HOGMASTER	937.6	
40	126	BRYCE BOAR BLAZERS	935.3	
41	146	NOTORIOUS P.I.G.	925.8	

Memphis in May International Festival
World Championship Cooking Contest
Grand Champions

YEAR	***GRAND CHAMPION***	**CATEGORY**
1978	Bessie Lou Cathey	Ribs
1979	Don Burdison	Ribs
1980	John Wills	Ribs
1981	John Wills	Ribs
1982	Martec Coaters	Shoulder
1983	Willingham's River City Rooters	Ribs
1984	Willingham's River City Rooters	Ribs
1985	**Holy Smokers Too**	Whole Hog
1986	Pig Iron Porkers	Whole Hog
1987	Cajun Country Cookers	Whole Hog
1988	**Holy Smokers Too**	Whole Hog
1989	Super Swine Sizzlers	Whole Hog
1990	Apple City Barbecue	Ribs
1991	David Cox Barbecue Team	Ribs
1992	Apple City Barbecue	Ribs
1993	The Other Side	Shoulder
1994*	Apple City Barbecue	Ribs
1995	Rebel Roaster Revue	Shoulder
1996	Pyropigmaniacs	Ribs
1997	Wildfire Gourmet Cooking Team	Shoulder
1998	The Other Team	Ribs
1999	Custom Cookers	Whole Hog
2000	Big Bob Gibson	Shoulder
2001	Jack's Old South	Whole Hog
2002	Pyropigmaniacs	Ribs
2003	Big Bob Gibson	Shoulder
2004	Jack's Old South	Whole Hog
2005	Gwatney Championship BBQ Team	Whole Hog

1994 was the last year Holy Smokers Too won whole hog and the end of our Decade of Dominance after winning the first time in 1985.

Preparing for Competition

As previously mentioned, the Memphis in May / Memphis BBQ Network sanctioned contests include both blind judging and on-site judging. Some teams have gone over the top with their team sites, including building a first-class restaurant motif within their allotted space. On the other hand, I've judged a contest where a competitor plopped his prepared shoulder on a door across two saw horses without seating for the on-site judge. He came in third out of 25, without the benefit of an upscale site. Fancy tablecloths and china plates should not have an influence on your judges. But, remember judges are people, too and subtle details can be the difference between a win or a loss. Whatever your budget or creativity allows, just make sure it is a clean and pleasurable experience for the judges.

Judging takes place in three distinct phases, each with different challenges. Blind samples are submitted in Styrofoam boxes during a fifteen minute window prior to on-site judging for that category. On-site judges go to contestant's sites beginning on the hour and at twenty minute intervals allowing for fifteen minutes at each team that is judged. The top three teams in each category will be notified and then a final round of four judges will score the top nine teams.

While all the showmanship is exciting during on-site judging, the blind judging is where the preliminary contest is either won or lost.

You will not be there to explain your process or what you were trying to achieve when your cooling BBQ, in its little white box, is placed side-to-side with 4 – 6 teams on the blind judges' table. Your meat will have to speak for itself. Most teams will wait until the last minute to submit their blind box hoping to keep their product as hot as possible. By the time it's checked in and the judges have the opportunity to judge your BBQ it will not be hot. At best, it will not be cold. If you submit sauce with your BBQ, the BBQ will be tasted, both with and without the sauce. We have found that it makes sense to put as much BBQ in the box as possible. This addresses two concerns: (1) the more meat mass, the warmer the product and (2) you never want a judge to feel slighted if he really likes your product and needs a second taste to decide if yours is the best BBQ on the table.

Before any of the judges taste your product, every judge at the table will observe the open boxes and score the entries on their appearance. So, if you dropped your box on the way over, they'll probably notice. After each judge has scored the entries on appearance, each judge will take a sample from each team box and place it on a plate before him. An experienced judge will have marked his plate off into sections to keep each team's BBQ separated and identified with the right box number. Then, the judges will consider the categories of flavor and tenderness. Finally, each team will be judged on overall impression. Again, this is where the judge can apply tenths of a point to separate teams with tie scores.

Pork shoulder blind boxes should be as full as reasonably possible to retain heat. Your box should contain a good mixture of meat pulled from deep into the shoulder as well as pieces containing barq. Make sure you include the most tender pieces of barq you can find. Tough and/or crunchy barq is not what a judge is looking for to give high scores. The pieces with the barq from the outside of the shoulder will have more flavor and show more smoke ring, but the deep pieces will have more moisture. Thumb-sized pieces hold heat and moisture well. In addition, the judges will pick the pork up in their hands to squeeze it and pull it apart as the first measure of tenderness and moisture content. This works best with large pieces. BBQ that's been pulled too thinly will dry out quickly. The same goes for Carolina BBQ hash, unless you've soaked it with sauce. You can sauce your BBQ in your blind box, but most judges want to taste your pork by itself and then sauce it for a second taste.

Myron Mixon of Jack's Old South is one of the most successful competitors because he knows how to cater to the regional judges. When competing in North Carolina he includes chopped shoulder with his pulled pork in both his shoulder and whole hog blind entries in case his judges like Carolina BBQ hash.

Ribs should have at least six ribs. I know there are only four judges scoring your ribs, but as I said earlier you want to make sure an interested judge has a second chance to score your ribs. Plus, if I'm the last judge to pull a rib from the box, I'll appreciate your giving me a choice of ribs to judge. Beauty is in the eye of the beholder. Teams will generally select the best ribs from the center of their best looking racks. If you have to pull one rib from each rack to get your winners, then do it. Remember, we're not saving good ribs to take home or to empress our friends who will come by later. I like to cut a *two-sided rib. A two-sided rib is created by cutting along the bone of the opposing ribs on both sides. This gives me a rib with double the meat of a typical single rib.* Then, place the ribs in a spooning or slightly overlapping arrangement so that the judges get to see both the surface and the smoke ring at their first glance. Eight ribs in the box are not too many.

Whole hog blind boxes are a real challenge. The official judging portions include the loin, shoulder and ham. Each portion should be placed so that it's easily identified and of ample quantity to serve four judges. Ham and shoulder portions are generally presented similar to the shoulder boxes. The tenderloin portion is best served in medallion style or large chunks to prevent it from drying out. There should be ample portions of center meat as well as smoke ring and barq. *If you are in N.C. you may want to include some finely chopped hash.* Jowl meat is not a required judging portion, but if a couple pieces of jowl meat find their way into your box, you may find an approving judge.

Blind boxes must be delivered to the judges' tent during the fifteen minute interval before on-site judging for that meat category. Judging times are usually set around 10:00, 11:00 and 12:00. If you are in line outside of the judging area when the window closes your entry will be accepted. Don't wait until the last minute to deliver your blind box. If you miss that window – your contest is over. For those of you who like to cut things close, you should have your blind box runner walk the route the previous day or first thing Saturday morning to get a feel for the time required as well as make sure you have the right location. There is nothing so pitiful as seeing a team member, Styrofoam box in hand, walking around the contest area at five minutes after the hour with that "Oh ?%&!" look on their face.

Contest Time Table

Planning your day during a contest is crucial. A great product that is ready after the turn in time may please your friends, but not the judges that will not see it. Below is a sample from an anal retentive team member who was far too intense and asked to leave. You can write your own.

Preliminary Rib Judging at 1:00 PM

6:30 AM - Start Fire. Place fire tray in center of grill. Stack three scoops of charcoal and light the fire. Desired temperature is 300° at 8:15 AM. Skin ribs.

6:45 AM – Apply Dry Rub. Ribs should be wrapped and allowed to come to room temperature.

8:15 AM – Ribs on Grill. Place ribs back-side down on grill. Arrange around fire so that none of the slabs are directly over the heat. Place big bone side of the slab toward the fire. This arrangement will leave an open square on grill surface directly above the charcoal. Put a hearty handful of hickory on the fire. Ribs should be on and the temperature at 275° - 300° at 8:15 AM.

9:15 AM – 1st Baste. Turn slabs on meat side down and baste the back side. Examine each to determine if the slabs closest to the fire are over cooking. Swap the outside slabs with those closer to the fire if need be. Turn the slabs back over, meat side up, and baste. Do this process rapidly to minimize lost cooking time. Add charcoal as necessary to maintain 275° temperature.

10:00 AM – 2nd Baste. Repeat above. Move slabs as necessary to maintain uniform doneness.

10:30 AM – 3rd Baste. Baste quickly to maintain heat.

11:00 AM – Wrap. Place individual slab in foil, meat side down. Baste back side then turn and amply baste top. Select 6 slabs for finals and place in empty cooler, NO ICE! Do not open the cooler until Finals Phase. The fire should be hot, 275°, during the first hour of wrap cook phase. The fire should taper down to around 200° at the layout phase.

12:30 PM – Layout Phase. Unwrap six – eight slabs for blind and 1st judge. Apply finishing rub and cider mist. This phase is critical because ribs can dry out in seconds if the fire is too hot. Apply mist as necessary to keep moist.

12:45 PM – Blind. Prepare blind then layout and mist remaining slabs for 2nd / 3rd judges.

Finals Judging at 4:00 PM

2:00 PM – Rebuild Fire. 275° temperature required at 2:30 PM

2:30 PM – Ribs on Grill. Retrieve the wrapped ribs from the cooler and place around fire.

3:40 PM – Layout. Unwrap, rub and mist as described above

On Site Presentations

While your team runner is delivering the blind box sample to the judges table, you should be finishing your preparation to receive the first of your three judges that will visit your area. Your space need not be fancy, but it should be clean and inviting to anyone who would come into your area and eat your BBQ.

Contests work hard to keep judges on time. Your judges should arrive at 10:00, 10:20, and 10:40 for your first category. This gives the judges fifteen minutes to hear your presentation and then five minutes to complete their scorecards and get to their next team. It's also your responsibility to keep track of your time and get them out of your area on time. Either the judge or an ambassador will signal it's time to go whether you have finished your presentation or not. If a judge arrives a little early or you have been pushed past your fifteen minutes by a talkative judge or slight accident in your area, it is O.K. to ask for a little more time. If you delay too long, the judge may tell you that you have less than the full fifteen minutes. This may or may not affect your score on presentation – it's the judge's prerogative.

If a judge presents himself to your area and you know your team has a negative relationship with this judge, it is appropriate to ask for a replacement judge. This also applies to a judge that has obviously been drinking alcohol during judging. This may be a problem for small contests, but it is your right to have your hard work judged fairly.

When you are ready for judging, with the appropriate number of team members present (MIM puts a limit of five team members in the area during judging), a team member should great the judge and welcome him into the area. You should check the judges score card to verify he is at the right team in the right time slot. It's not unusual for judges to get confused if they are new to a particular contest or have not been there early enough to walk the grounds. Truth be told, some contests don't do a very good job of identifying team locations or provide area maps. And then again, not all teams have a large team banner that's easily identified.

It's not necessary, but a nice touch is to identify each members responsibilities during this contest as well as where they are from.

The following is just a suggestion on the process of presenting your product to the judge – you can use it, change it, or throw it out. It's up to you. I have found it works well to cover all the required information and is followed by many of the teams I have judged as well as the Holy Smokers Too.

Take the judge over to your smoker and explain your cooking process:

(1) Discuss your raw product – where did you get you meat. How much did it weigh? How did you prepare your product? How was it butchered and trimmed? Did you brine or marinate? How long? What type of dry rub? How long was the dry rub allowed to stay on the meat before cooking?

(2) Describe your cooker and the cooking process – How was it made? What type of heat source do you use? Wood, charcoal, pellets, etc. How much did you use during the process. What temperature did you cook you meat and how long? Did you baste or otherwise tend to your meat during the process?

Show the judge your product while it's on the smoker so he gets the overall impression of the smoker and your product. When you have completed the process portion of the presentation, ask the judge if there are any questions and invite the judge to have a seat for the tasting. If you have at least a two man team, this is a good time to tell the judge more about your team. While your teammate is preparing the samples for tasting, this is a good time to talk more about your dry rub and sauces and allow the judge to taste each.

Some teams will bring a rack of ribs or whole shoulder to the table and let the judge pull or select the meat he wants to taste. This is a bold move, but somewhat risky. You may want to pull a couple of ribs you want judged or the sample of your pork shoulder to make sure it's your best product. Both whole hog and shoulders have a lot of fat and the last thing you want a judge to put in his mouth is a large glob of pig fat. Remember, this judge is judging three or more teams and just needs a taste to judge your product, so give him the best possible taste experience.

Presentation Dialogue

As your first judge approaches for rib judging at 12:00, your presentation might go something like this:

Billy Bob Billy: Welcome to Billy Bob's BBQ. I'm Billy Bob and these are my team members Larry, Curly and Moe. Let's check your scorecard and make sure you have the right team. Yep, it says team one is Billy Bob's BBQ @ 12:00.

Judge: Here's your judge evaluation form.

BB: Thanks. Curly is going to take you over and explain our cooking process for you.

Curly: Hi judge, I'm the Rib Master for the team, so I'm going to go through our cooking process with you. We want to keep you on time, but feel free to ask questions any time.

Judge: Great, go ahead.

Curly: We started preparing these ribs late last night about 10:00 P.M. We brought ten racks of ribs we bought at Sam's Club before leaving Knoxville. We were looking for pork loin (baby back) ribs in the 1.75 to 2.00 pound size. We feel these come from younger pigs and are less fatty and more tender than from larger pigs. First, we trimmed any excess fat and removed the membrane from the bone side of the ribs. We feel removing the membrane allows better penetration of our rubs and baste. Once we liberally applied our rub to the ribs we wrapped them in plastic wrap and placed them in our coolers overnight.

Curly: Any questions so far?

Judge: Nope. It all sounds good so far.

Curly: This morning, about 6:00 A.M. we pulled the ribs from the cooler and spread them out on our preparation tables to allow them to warm up. We sprayed them with a little apple juice before putting them on the smoker at 7:00 A.M.

Judge: When did you light the smoker?

Curly: I'm glad you asked about the smoker. That's Moe's pride and joy.

Moe: Our smoker was built by Lange Smokers in Georgia. It looks like a propane tank made of 1/8 inch rolled steel with an offset fire box. One of its unique features is a steel plate that runs the length of the cooking chamber forcing the smoke to travel down to one end and back to exit the smokestack. This allows us to maintain a consistent temperature across the entire smoker. While Curly was prepping the ribs this morning I lit the fire about 6:00 A.M. I brought the smoker to an initial temperature of 350° to make sure the smoker was ready for the ribs.

I bring it to a high temperature because I know we will lose heat when we open the doors and place our ribs inside. We then maintained a cooking temperature of 225° thru out the four hour cooking process. We used oak logs for heat and added soaked apple wood chips every thirty minutes, or when we no longer saw smoke coming from the stack. We used three bags of chips during the first two hours.

Judge: You only smoked the ribs for the first two hours?

Curly: That's right. We smoke the ribs for two hours, basting at 30 minutes and 90 minutes. Then we wrap the ribs individually with ¼ cup of basting sauce and allow the ribs to rehydrate and tenderize for one hour. At the end of the third hour we remove the ribs from the foil and place them back on the smoker for the final hour. Thirty minutes before the end of the four hour process we glaze our ribs with our BBQ sauce. We glaze them again fifteen minutes later and will glaze them one more time before we serve them to you. Are you ready?

Judge: I'm ready.

Moe takes the judge over to the table where BB is waiting while Curly prepares the ribs for presentation. As the judge sits down, BB begins his presentation.

BB: Well, did Curly and Moe answer all of your questions?

Judge: I think so. You can go ahead with your presentation.

BB: First, I want to thank you for taking your time to come out and judge us today. Without volunteer judges like you we would not have a contest. Let me draw your attention to our area. We had about fifty old and new friends here last night, so we had to get up early and clean the place up for you. It's not just for you, because this weekend our site is our home-away-from-home. You will also notice that all team members who come in contact with your food are wearing gloves. We even got Larry to shave this morning. We wanted it to be clean for our guests and the judges today.

> *Area and Personal Appearance only count for 2%, but you want to make sure you get credit for your hard work.*

BB: Next, let me tell you about our dry rub and baste. We use a combination of black pepper, paprika, Chile powder, garlic powder, celery seed, dry mustard and cumin for our dry rub. A lot of teams use similar rubs, but we think our cumin adds a great smoky flavor. There is no salt in our rub because we don't want to draw moisture out of the ribs. There is also no sugar because sugar tends to burn and form a caramelized coating we don't like. I also like to add a sprinkling of sweet basil on the ribs just before we place them on the smoker. Here, have a taste.

Judge: That's good, but a little hot for my taste.

BB: A lot of people say that when they taste it out of the shaker, but during the cooking process the heat dissipates to a reasonable level that complements the smoked pork. I think you'll like it. In addition, we build layers of flavor during the cooking process with our basting liquid. We use mixture of our BBQ sauce, Wickers marinade, lemon juice and apple cider for our rib baste. The lemon juice and cider are a great acid balance to the sweetness of the BBQ sauce.

BB: We also use our BBQ sauce to glaze the ribs when we serve "wet" ribs. The BBQ sauce is a modified version of the Holy Smokers Too sauce my old team used. That sauce is based on a recipe that won the 1946 Mid-South Fair in Memphis. The sauce has 23 ingredients and 16 spices. Here, taste it for yourself.

The first thing you'll taste on the tip of your tongue is the sweetness from the brown sugar. Next, your mouth will form a slight pucker as the citrus rolls over the middle of your tongue. And finally, there will be a little bit of heat as it slides down your throat, just to remind you that this is true Southern BBQ sauce.

BB: Here comes Larry with your ribs. Man, they sure do look good! Curly sure put a beautiful glaze on those ribs today. Judge, I want you to take note of the smoke ring on the sides of those ribs. Moe really pours the smoke on in the first two hours to produce that fantastic smoke ring. As you pull the meat from the bone, you'll notice the bared bone will begin to turn white or "marbleize." That shows that the rib is fully cooked. A good portion of the rib cooks from the inside-out from heat radiating down the bones. There should be a little resistance as the meat is pulled from the bone, but we believe it should come away and leave a clean bone.

BB: If we have done our job today you should taste the smoky rib meat and the flavors from the dry rub and the glazing sauce. The spices and sauce should complement the pork, but not overwhelm the rib. You can cover a lot of cooking mistakes with a heavy sweet sauce, but then you don't taste the rib.

Judge: Tastes great. How do you control the smoke ring?

BB: Ha! Ha! That's really a trick question. The depth of the smoke ring depends on multiple factors including cooking temperature, the amount of smoke generated, time of exposure, as well as how tender and receptive the ribs are. If we could control that we would win every contest. We just follow our process we have fine tuned over our years of trying to produce the best possible product in each contest. Any other questions?

Judge: I don't think so.

BB: Great, that brings me to the end of our presentation. Judge we want to thank you again for volunteering to make this contest work. We'd like to invite you back after your judging to enjoy our BBQ and a cold one. This is a very competitive contest with a lot of great teams. It will take 10s to get into the finals and we hope we have done a good enough job today to earn some 10s on your scorecard. Thanks again.

> *As you go through your presentation you want to follow the proven sales technique: Tell 'em what they are going to taste, let 'em taste it, and then tell 'em what they tasted. And last, but not least, ask for the order (scores of 10 in this case).*

You will need to practice your own presentation for timing and flow. Timing is critical. Even old pros need to practice. At the first contest in a long time after our decade-of-dominance period that we made the finals we were a little rusty. We spent so much time at the smoker telling about the smoker and the cooking process that we ran out of time trying to pull the meat for tasting for the finals judges. The result: 5s in flavor, appearance, and tenderness. So, we locked in an early third place.

If you are uncomfortable with your presentation your tenseness will be noticed by your judge and neither you nor the judge will have a comfortable, relaxed experience. It is possible your score will reflect the tension. Remember, too, that judges can be nervous if they are a new judge. The more comfortable you can make them the better your overall impression scores will be.

You did start this to have fun, right!

Practice!   Practice!   Practice!

Judging During Finals

On-site finals judging presentations don't vary from your preliminary judging except you are presenting to four judges at one time. This will require some logistical focus, but your presentation should be just as detailed, if not more so, as your earlier presentation.

If you happened to have made the finals in more than one category you may want to change your jokes and leave out some of the team history the second or third time the finals judges visit your area. We should all be so lucky to have that problem.

One additional challenge is to "hold" your finished product from preliminary judging in the morning until finals judging in the afternoon. It's fairly easy to set a shoulder or two aside and wrapped in a warm (not hot) section of the smoker so it will not cook any further but will stay hot without drying out. For whole hog, most team will pull preliminary meat from the back side of the hog and save the front for the finals judges. This maintains the hog's best appearance for the judges prior to pulling meat for the finals judges. Ribs can easily be started at separate times for preliminary judges and finals judges. A sample rib schedule follows this section.

Team exuberance must also be managed. In a Missouri contest years ago we were next to a team of good friends who were always a major competitor. We both made the finals in whole hog. The team next to us took great pride in taunting the Holy Smokers and celebrating their making the finals. This celebration, along with alcohol consumption in large quantities went on for the whole five hours prior to the finals judges starting their rounds. When the finals judges completed our judging, we watched with great interest as they went next door to judge the other whole hog team. Their presentation was flawless and impressive right down to serving their hog samples. As their head cook approached the table the partying and beer drinking exacted its revenge. He fell, face first across the table, with pork flying in every direction. The judges sat stunned until we heard a team member say, "Well, that concludes our presentation. The only thing I can say is congratulations Holy Smokers Too." So, beware, it's not over until it's over. Save your serious partying for the award ceremony, or at least, after the last judge leaves your area.

Billy Bob Billy's
Message to Friends and Family

The day will come when my body will lie upon a white sheet neatly tucked under four corners of a mattress located in a hospital busily occupied by the living and the dying. At a certain moment a doctor will determine that my brain has ceased to function and that, for all intents and purposes, my life has ended.

When that happens, do not attempt to instill artificial life into my body. Save the medical miracles for those who have not enjoyed the full life and love of family that has blessed me. And don't call this my deathbed. Let it be called the Bed of Life, and let my body be taken from it to help others improve their lives.

Give my sight to the man who has never seen a sunrise, a baby's face, or love in the eyes of a woman. Give my heart to a person whose own heart has caused nothing but endless days of pain. Give my blood to the teenager who was pulled from the wreckage of his car, so that he might live to see his grandchildren play. Give my kidneys to one whose hours of every day are spent tied to a machine. Take my bones, every muscle, every fiber and nerve in my body and find a way to make a crippled child walk.

Explore every corner of my brain to further the understanding of why we do what we do so that others may lead happier lives. Take my cells, if necessary, and let them grow so that, someday, a speechless boy will shout at the crack of a bat and a deaf girl will hear the sound of a kitten purring.

Burn what is left of me and place it in a vessel that can be cast into the Mississippi River at Tom Lee Park, where I have spent a week of my life every year, for over 25 years, smokin' BBQ with friends and family. This will be my last trip to New Orleans.

If you want to bury something, let it be my faults, my weaknesses and the mistakes I have made. If, by chance, you wish to remember me, do it with a kind deed or a word to someone who needs you. Give a smile to a stranger; what you will receive in return may amaze you. Pat a child on the head; if you're lucky, some of their energy will rejuvenate your spirit. Hug someone at sunset while you gaze into your future together. If you do all I have asked. I will live forever.

I have paraphrased an article I read over 30 years ago. The author was unknown, but I am sure I am one of many who are better for having read it. I pass my version on to you, with the hopes that you will take it to heart and write one of your own to share with the people in your life that are important to you.

When your time comes, the good Lord will be there to look at your life and say, "show me your product." When Louis Fineberg was fatally injured, there was nothing anyone could do for him; but we all can do something for somebody, every day, if we put our hearts in it.

Billy Bob Billy

Reference, Resources and Recommendations

Books

Peace, Love and Barbecue by Mike Mills and Amy Mills Tunnicliffe

Mike and his daughter take you across America to meet some of the legends of BBQ lore. A great read on a cold night with a roaring fire and your favorite beverage.

How to Grill by Steve Raichlen
Also known as *The BBQ Bible*

Probably the best illustrated guide to cooking on the grill that's out there today. This book has detailed illustrations of the details in food preparation and grilling. It's a great resource when you're facing down a piece of meat you have never heard of or a technique that's foreign to you.

Sublime Smoke by Cheryl and Bill Jamison

It would take you a lifetime to compile and test the gamut of recipes the Jamisons have included in this book. When you're feeling like something new, but your creative juices are low, this book will lead you to new territory.

The Meat We Eat by Romans, Costello, Carlson & Jones

Pure textbook used by many universities. If you really want to know how things work and where your meat comes from, this is your book. Also serves as a good door stop and cure for insomnia.

Web Sites

www.HolySmokersToo.com A great site for me to sell you more stuff as well as for you to pick up more tips and techniques from monthly postings.

www.outreach.utk.edu University of Tennessee site to register for Billy Bob Billy's Championship BBQ Cooking Class.

www.thesmokering.com The BBQ networking site for everything to do with BBQ and those that do.

www.mbnbbq.com Official site of the Memphis BBQ Network. The sanctioning body that picked up the challenge when Memphis in May decided there was more money in the festival business than in sanctioning contests.

www.memphisinmay.org Official site of Memphis in May Festival

www.kcbs.us Official site of the Kansas City BBQ Society

www.pigroast.com Home for Lang Smokers. Probably the best bang for your buck in competitive smokers.

www.nbbqa.org Home for the National BBQ Association. **The site for people in the business of BBQ.**

The internet allows you to "Google" on BBQ and find unlimited resources for BBQ products and information. I hope the above sites give you a jump start.

INDEX